CAMBRIDGE GEOLOGICAL SERIES

AGRICULTURAL GEOLOGY

AGRICULTURAL GEOLOGY

BY

R. H. RASTALL, M.A.

Late Fellow of Christ's College and Demonstrator of Geology
in the University of Cambridge

Cambridge:
at the University Press
1916

CAMBRIDGE
UNIVERSITY PRESS

University Printing House, Cambridge CB2 8BS, United Kingdom

Published in the United States of America by Cambridge University Press, New York

Cambridge University Press is part of the University of Cambridge.

It furthers the University's mission by disseminating knowledge in the pursuit of education, learning and research at the highest international levels of excellence.

www.cambridge.org
Information on this title: www.cambridge.org/9781107673090

© Cambridge University Press 1916

First published 1916
First paperback edition 2014

A catalogue record for this publication is available from the British Library

ISBN 978-1-107-67309-0 Paperback

PREFACE

OF late years a large and increasing amount of attention has been paid to the practical applications of geological science. The present book has been written with the object of supplying a concise account of those parts of geology which are of direct interest to the agriculturist. The study of the soil naturally forms the most important part of the book, and this has been treated as much as possible from a purely geological standpoint: the chemical and physical properties of soils are dealt with exhaustively in many books. It is to be regretted that considerations of space did not allow of a fuller discussion of the modern mechanical methods of soil-analysis that have proved so valuable in the hands of workers at Rothamsted, Cambridge and elsewhere. The later chapters of the book contain a summary of the distribution of the rock-formations of the British Isles, and the characters of the soils yielded by them. Much difficulty was experienced in obtaining the necessary information, which is still far from complete. The facts had to be collected from many scattered sources, especially from the memoirs of the Geological Survey, the *Journal of the Royal Agricultural Society*, county histories and geographies and many other publications. Much valuable information was also derived from Mr A. D. Hall's *Pilgrimage of British Farming*. I must here express my gratitude to those gentlemen, too numerous to mention individually, who afforded me information in reply to direct enquiries.

For much help and encouragement in the preparation of the book I am indebted to Mr K. J. J. Mackenzie, Reader in Agriculture at Cambridge, and to Dr F. H. A. Marshall,

Fellow and Tutor of Christ's College, Cambridge. The latter was so kind as to undertake the preparation of the last chapter of the book, on the geological history of farm animals. Permission to reproduce the figures illustrating this chapter was kindly given by Professor J. C. Ewart, Messrs George Allen and Unwin, Ltd., Messrs Methuen and Co., and the Highland and Agricultural Society. Professor J. C. Ewart and Professor T. M^cK. Hughes also supplied some information incorporated in this chapter.

Many of the text-figures in the book were drawn for me by Dr J. E. Marr, to whom my thanks are due. Finally, I am much indebted to Mr W. H. Wilcockson, who read all the proofs and assisted in the preparation of the index.

R. H. R.

April 1916.

CONTENTS

ILLUSTRATIONS

CHAPTER I

INTRODUCTION. MINERALS AND ROCKS

Definition of a rock. The principal object of the agricultural geologist is the study of the soil, but since all soils are formed either directly or indirectly from rocks, it is essential for a proper understanding of the subject to gain a preliminary acquaintance with the composition and properties of the rocks themselves, which are the ultimate source of the materials of the soil. This material has as a rule undergone many vicissitudes before attaining the condition in which it is found at the present time. All geological processes work in cycles of change, which commonly have no definite beginning and no definite end; hence it becomes necessary to adopt some more or less arbitrary starting-point in any discussion of the processes which are continuously taking place. Rocks are continually being formed and again destroyed; they are undergoing ceaseless change, but these changes are generally very slow, and it is often possible to find a rock in what is technically known as a *fresh* condition, this term being used to signify the fact that it has undergone little or no visible alteration since its consolidation into its present form. It is convenient therefore to begin with a study of fresh rocks, from the mineralogical and chemical standpoint, and to leave the investigation of subsequent changes to a later stage.

To give an accurate scientific definition of a rock is somewhat more difficult than would appear at first sight. The term, as popularly applied, generally connotes the ideas of solidity and hardness, but this criterion soon breaks down when applied in practice, since rocks vary greatly in hardness and every transition may be found from the hardest granites and quartzites to forms which are actually quite unconsolidated and incoherent. Even among the older strata composing the earth's crust and at considerable depths below the surface, there may be found beds of loose sand and other materials, differing little from the modern deposits of the sea-shore or the superficial accumulations of the land-surfaces. To the geologist all these ancient deposits are rocks, whatever their state of aggregation, and in passing upwards from older to newer deposits it is impossible to fix on any definite horizon at which rocks cease and surface accumulations begin; in fact under certain conditions the most recent formations may be as solid and hard as anything which has been formed in the past. From this point of view rocks may be said to include all the solid materials forming the crust of the earth, so far as it is accessible to observation.

Perhaps however the most satisfactory definition of a rock is *an aggregate of mineral particles*. This definition makes no assumption as to the composition or state of aggregation of the particles and hence may be taken to include the accumulations of all ages and of all degrees of solidification and alteration, without regard to their manner of formation. As will appear more fully in a later section, in some cases rocks are formed by consolidation from a state of fusion, while in other cases they originate at or near the surface, under normal conditions of temperature and pressure, through the action of the ordinary geological agents, but the above definition holds in either case.

As will be explained more fully in a later section, rocks may be divided into two well-marked classes, differing fundamentally in their mode of origin. These are: (*a*) the *igneous* rocks, formed by consolidation from a state of fusion—an example of this class is afforded by the lava emitted during a volcanic eruption; (*b*) the *sedimentary* rocks, formed on the earth's surface or in water by the operations of ordinary geological agents. These are built up either directly or indirectly from the materials of pre-existing rocks and the ultimate source of the materials is to be sought in the igneous group. Such are the sandstones, clays and other common types which form the land-surface of a great part of the world. Under the heading of *metamorphic* rocks are included all those varieties which have undergone since their deposition such changes that they have assumed new characters; in some cases these changes are so far-reaching that it is difficult or impossible to decide whether the rocks were originally igneous or sedimentary.

Chemical composition of rocks. The rocks composing the earth's crust must obviously contain all the known elements, some eighty in number, with the possible exception of a few existing only in the atmosphere, but the number of elements which occur in large proportion is very small. According to the most recent computations only eight elements occur in amounts exceeding 1 per cent. of the whole, namely, oxygen, silicon, aluminium, iron, calcium, magnesium, sodium and potassium. These, together with carbon, sulphur, phosphorus, titanium, hydrogen and chlorine form all the common rock-forming minerals.

The results of analysis of rocks and minerals are generally calculated not as elements, but as oxides, and when stated in this way the composition of the earth's crust, or lithosphere, is approximately as follows[1]:

	Per cent.
Silica, SiO_2	59·85
Alumina, Al_2O_3	14·87
Ferric oxide, Fe_2O_3	2·63
Ferrous oxide, FeO	3·35
Magnesia, MgO	3·77
Lime, CaO	4·81
Soda, Na_2O	3·29
Potash, K_2O	3·02
Water, H_2O	2·05
Titanium dioxide, TiO_2	·73
Carbon dioxide, CO_2	·70
Phosphorus pentoxide, P_2O_5	·25
Sulphur, S	·10
Chlorine, Cl	·06
	99·48

[1] Clarke, "The Data of Geochemistry," 2nd edition, *Bulletin* 491, *United States Geological Survey*, 1911, p. 32.

From these figures it will be seen that the percentage of certain substances, such as carbon, phosphorus, sulphur and chlorine, is very small, but they are all elements of much scientific and practical importance and are therefore included. The proportion of the valuable metals, gold, silver, copper, tin, lead, etc., is insignificant as compared with the substances above enumerated. The element titanium, though very widely distributed, is of no practical importance.

Rock-forming minerals. It has already been stated that the constituents of rocks are minerals, and it now becomes necessary to understand the meaning of this term. A *mineral* may be defined as a naturally occurring inorganic substance, possessing definite physical properties and in most cases a definite crystal-form, though certain true minerals occur in an amorphous or non-crystalline condition. In many instances the composition of minerals can be expressed by simple chemical formulae, since they are the crystalline forms of pure chemical compounds (e.g. quartz, SiO_2; calcite, $CaCO_3$; magnetite, Fe_3O_4), or even as elements (e.g. diamond, C; sulphur, S, etc.). But many of the most important rock-forming minerals are not pure chemical compounds; they are to be regarded rather as mixtures of various compounds possessing the property of isomorphism; in other words they are *mixed crystals*. Under this heading come most of the silicates, a group of minerals of the highest importance as constituents of rocks and soils.

The total number of mineral species which have been recognized by systematic mineralogists is enormous and ever increasing, but fortunately only a very small proportion of these are of practical importance to the geologist. It is possible to draw up a list of less than twenty minerals, which together constitute 99 per cent. of the whole visible crust of the earth. For most purposes the rest may be disregarded.

The following list comprises the varieties which are of most common occurrence as rock-builders; some of them can be formed only by crystallization from a fused state, while others can only originate at the ordinary temperature and pressure, that is, as sedimentary deposits. A few are common to both groups.

Quartz	Olivine	Rock-salt
Felspar	Magnetite	Gypsum
Mica	Iron pyrites	Apatite
Hornblende	Calcite	Garnet
Augite	Dolomite	

The student is recommended to make himself thoroughly familiar with the chemical composition and physical characters of these minerals; reading should be supplemented by actual examination of well-selected specimens, until the different varieties can be readily recognized by inspection, or by an application of the simple tests described in any text-book of mineralogy[1].

Quartz is the commonest of the crystalline forms of silicon dioxide or silica, SiO_2. It crystallizes in the hexagonal system, most usually in the form of a six-sided prism surmounted by a six-sided pyramid. It is also frequently found as irregular masses, or as aggregates of small crystals and crystalline grains. It possesses no regular cleavage, but breaks with an irregular and often curved fracture. The colour is very variable; some specimens are perfectly colourless, clear and transparent, while other varieties are opaque and milky in appearance, or show various shades of brown, yellow, pink, or purple; the variations in colour are due to the presence of minute quantities of some impurity. Quartz is very hard, scratching glass easily, and its specific gravity is about 2·65. It is totally unaffected by any acids, except hydrofluoric acid, and is one of the most stable and resistant of all minerals.

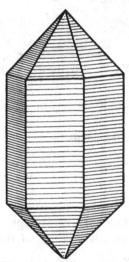

Fig. 1. A simple crystal of quartz, consisting of a hexagonal prism, terminated by hexagonal pyramids.

Note. Silica occurs widely in nature in other forms, differing from quartz in their physical properties. They are either obscurely crystalline or amorphous, forming such substances

[1] Hatch, *Mineralogy*, 4th edition, London, 1912.

as opal, chalcedony, jasper and flint. Most of these possess a lower specific gravity than quartz (opal, 2·1) and are more easily attacked by acids and especially by alkalies.

Felspar. Under this heading are comprised a large number of minerals to which systematic mineralogists apply different names. Their chemical composi-

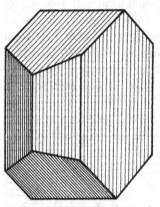

tion is rather complex, but all contain silica, alumina, and either potash, soda or lime, or some mixture of the three latter bases. For convenience they may be divided into the potash-felspar or *orthoclase* group, in which potash is dominant, and the soda-lime or *plagioclase* group, containing soda or lime, or a mixture of the two. These two groups differ slightly in physical properties, though their general appearance is very similar.

Fig. 2. A typical simple crystal of orthoclase felspar.

The chemical composition of orthoclase may be expressed by either of the formulae, $KAlSi_3O_8$, or $K_2O . Al_2O_3 . 6SiO_2$. The latter form is to be preferred. This corresponds to

				Per cent.
Silica	...	...	...	64·7
Alumina...	...	...		18·4
Potash	...	...	...	16·9
				100·0

Hence the percentage of potash is seen to be high, and in fact orthoclase is one of the most important sources of potassium in soils.

The felspars of the plagioclase group may be regarded as mixtures in any proportion of the two ideal compounds: albite, $Na_2O . Al_2O_3 . 6SiO_2$, and anorthite, $CaO . Al_2O_3 . 2SiO_2$.

		Albite	Anorthite
Silica	...	68·7	43·2
Alumina	...	19·5	36·7
Soda	...	11·8	—
Lime	...	—	20·1
		100·0	100·0

Special names have been applied to many intermediate varieties, but for our purpose it is unnecessary to enumerate these.

Orthoclase crystallizes in the oblique, and plagioclase in the anorthic system. The general forms of the crystals are however very similar. All felspars possess two very perfect cleavages, which are at right angles in orthoclase and nearly so in plagioclase. The cleavage faces generally show a conspicuous pearly lustre. The colour is variable; generally white or grey, or a pale shade of pink or red; green varieties are less common. Occasionally a peculiar bluish iridescence is noticeable. The mineral is hard, though softer than quartz, and the specific gravity varies from 2·6 to 2·7, the varieties rich in lime being the heavier.

The felspars show very conspicuously the peculiar structures known as twinning. For an explanation of this term a text-book of mineralogy must be consulted. It will suffice here to say that this phenomenon affords a ready means of distinction between orthoclase and plagioclase, since most of the faces of the latter mineral are marked by fine parallel striations due to twinning of a kind which cannot occur in orthoclase. The presence of these striations affords an infallible diagnosis for plagioclase.

Mica. The micas constitute a group of minerals of rather complex chemical composition, possessing well-marked physical characters which render their identification easy. They are silicates containing alumina and one or more of the following bases: potash, soda, iron or magnesia. Those varieties rich in iron are deeply coloured, usually brown, and may be designated *biotite*, while the varieties without iron are pale or colourless and may be placed under the heading *muscovite*. All micas contain more or less hydrogen, which probably exists in combination as the radical OH. The composition of an ideal muscovite may be provisionally represented by the formula $H_2O . K_2O . 2Al_2O_3 . 4SiO_2$. The constitution of biotite is even more complex than this.

All the micas possess one very perfect cleavage and show a strong tendency to form thin flakes with a brilliant metallic lustre. Muscovite is a remarkably stable mineral and is not affected by agents of decomposition, being therefore abundant in sediments as well as in igneous rocks, whereas biotite is easily decomposed by weathering and hence is characteristic of fresh rocks, both igneous and metamorphic.

Hornblende is a silicate of magnesia, iron and lime in varying proportions; some varieties also contain soda. It usually forms prismatic or needle-shaped crystals with an approximately hexagonal cross section. There are two good cleavages, which cross at angles of 124° and 56°. Most varieties of hornblende are black in colour, occasionally green, with a semi-metallic or somewhat resinous lustre. The mineral is hard and heavy (specific gravity about 3·1). It is stable and not attacked by acids. Hornblende is common in igneous and metamorphic rocks, but only occurs in sediments as derived grains.

Augite is very similar to hornblende in most of its properties, the chemical composition being almost identical. The crystals differ somewhat in form, usually showing eight prism faces instead of six, and the two cleavages are nearly at right angles. The colour and lustre are similar, but augite is heavier than hornblende, having a specific gravity of about 3·3. It is common in igneous rocks and is somewhat more liable to decomposition than hornblende, being therefore less abundant in sediments.

Olivine is a silicate of iron and magnesia; it may be regarded as a mixture of the two compounds Mg_2SiO_4 and Fe_2SiO_4. It occurs most commonly as crystalline aggregates and rounded grains, of a yellowish green colour. It shows little or no cleavage and a glassy lustre. Olivine is unstable and is only found in the igneous rocks.

Magnetite consists of the magnetic oxide of iron, Fe_3O_4. It forms well-defined crystals which are generally octahedral in form, or it may occur massive. It is black in colour, with a brilliant metallic lustre, and the density is high (about 5·0). Magnetite is specially common in igneous rocks, and occurs also in sediments to a considerable extent.

Iron pyrites, disulphide of iron, FeS_2, is an easily recognizable mineral with metallic lustre and the colour of brass; it is very hard and heavy, and most commonly occurs in the form of cubes. It is specially characteristic of sedimentary rocks of a muddy nature, which have been subjected to a certain amount of metamorphism.

Calcite is the more important of the two crystalline forms of calcium carbonate, $CaCO_3$. Its crystalline forms are very numerous and variable, all being referable to the rhombohedral subdivision of the hexagonal system. It possesses three very perfect cleavages arranged in the form of a rhombohedron. Cleavage faces show a strong pearly lustre. The colour is variable, but generally pale; some varieties are clear and colourless (Iceland spar), while others are milky white, yellowish or pink. Calcite is soft, being easily scratched with a knife, and it effervesces readily with dilute hydrochloric acid. Specific gravity 2·72.

Dolomite is a double carbonate of calcium and magnesium, $CaMg(CO_3)_2$. Its crystalline forms are similar to those of calcite, but simpler, and the crystals often show curved faces. The cleavages also are similar to those of calcite. The colour is generally creamy white, yellowish or pale brown, colourless varieties not being very common. Dolomite has the same hardness as calcite, but does not effervesce readily unless the acid is warmed. Specific gravity 2·85.

Rock-salt, sodium chloride, NaCl, is sometimes found as cubic crystals, but is more common in the massive form, in layers and beds in stratified rocks. It has no cleavage, and is very soft, being easily scratched by the finger-nail. It is readily soluble in water, and can be identified by the taste. When pure it is colourless, but is generally stained red or brown by an admixture of fine clay.

Gypsum is crystallized calcium sulphate, $CaSO_4 . 2H_2O$. It occurs either in oblique crystals disseminated through beds of clay, or in a fibrous massive form; the latter is often called *satin spar*. Gypsum possesses one very perfect cleavage and two less well-developed. It is even softer than rock-salt, but is only slightly soluble in water. It does not effervesce with acids.

Apatite is the crystalline form of calcium phosphate. It occurs very commonly in the igneous rocks in small hexagonal crystals. It is generally colourless, sometimes green or brown when in large crystals, fairly hard and heavy (specific gravity 3·2). Apatite is of great practical importance as the ultimate source of the phosphoric acid of the soil.

Garnet is a silicate, usually containing several metallic radicals, of which alumina is often present in large proportion. It crystallizes in complex forms belonging to the cubic system, and varies in colour from pink to brown and black. It is hard and possesses a brilliant lustre, with high refractive index; hence it is sometimes employed as a gem-stone. Garnet is most characteristic of metamorphic rocks, though it also occurs in those of purely igneous origin. It is also a common derived constituent of sands.

Besides the minerals enumerated above, many others are common as constituents of igneous and metamorphic rocks, of sediments and of the soils derived from them. Most of them are of little or no practical importance, and their study belongs rather to the province of mineralogy than of geology, and by the agriculturist at any rate they may safely be neglected. Some of the more important will be mentioned incidentally at a later stage when dealing with the constituents of soils.

Classification of rocks. Although the exact nature of the geological processes which took place in the earliest stages of the earth's history is still for the most part a matter of speculation, there can be no doubt that the earliest rocks were formed at a high temperature, and they must have resembled, at any rate in their mineralogical composition, those which we know to have solidified at later periods from a state of fusion. It is a safe inference to conclude that the material of all the sedimentary rocks was derived either directly or indirectly from this primitive crust, and perhaps to some extent also from the primitive atmosphere. Disregarding for the moment this latter possibility, it may be said that the minerals of the sediments are derived from igneous rocks, and therefore the study of rocks logically begins with this group, although at the present time they occupy much less of the earth's surface, and are of less importance as soil-formers.

The igneous rocks comprise all those masses which have solidified from a state of fusion, either at a greater or less depth within the earth's crust, or actually on the surface, as in the case of lavas. They are for the most part composed of an aggregate of crystals and grains of different minerals mixed in different proportions, e.g. quartz, felspar, mica, hornblende, olivine, etc. The deep-seated rocks are always completely crystalline, but in some cases where the molten material found itself at or near the surface, cooling was so rapid that crystals of definite minerals could not form, and the whole solidified as a homogeneous mass, which is known as a glass.

The second great group, known as the sedimentary, aqueous or stratified rocks, on the other hand, are formed at the ordinary surface temperature as a result of the geological processes which may be seen in operation around us at any time. The nature of these processes will be fully discussed in a later chapter; it must suffice here to say that as a result of the slow destruction of the land by water, ice, wind, etc., masses of material are loosened, transported by various agencies and piled up into masses, which may eventually become consolidated into hard rocks, or may remain loose and incoherent for a long period. The term *stratified* as above employed connotes the fact that these masses of sediment commonly occur in layers, or strata, which are at first generally horizontal, but may eventually be tilted at any angle. It is clear that the sedimentary rocks are formed from the materials of pre-existing rocks, with occasionally, as will be explained later, addition of matter of organic origin. Even this latter however is found on ultimate analysis to be derived from rocks, or from the atmosphere. The term *aqueous* is frequently used as synonymous with sedimentary, but it is not satisfactory, since there are important groups of sediments in whose formation water plays no part.

The third great group is that of the metamorphic rocks; the meaning of this term has already been briefly explained and all consideration of this group as a whole may be deferred for the present.

Rock-structures. By the term rock-structures is here understood the general features determining the forms of rock-masses on a large scale, as well as the disposition of their divisional planes and other surfaces of discontinuity, actual or potential. Rock-structures are the characters which are conspicuous during an examination of rocks in the field and they can often be distinguished at considerable distances, as for example in a general view of a cliff or of a mountain. The smaller features of a rock, necessitating observation at close range, are here separated under the heading of *rock-textures*. The most cursory examination of any section, natural or artificial, where rock-masses are exposed to view, will generally disclose more or less of a parallel arrangement; it is only in the case of some igneous rocks that this is entirely absent, and even these nearly always tend to break up into blocks bounded by plane or curved surfaces. Completely homogeneous masses measuring more than a few feet are rare.

In accordance with their widely differing mode of origin there is a good deal of difference in the structures that characterize the igneous and sedimentary rocks respectively, necessitating to a certain extent separate treatment for each group. The metamorphic rocks also possess structures peculiar to themselves, which are direct results of the metamorphic processes.

Forms assumed by igneous rock-masses. The forms assumed by masses of molten lava poured out by volcanoes will evidently depend on two factors, namely, the degree of liquidity of the lava, and the form of the surface over which it flows. A liquid lava will obviously spread out into a thinner sheet and flow further than a viscous one, under the same conditions, but the dominant factors are the form and slope of the surface of the ground over which they flow; since these may vary indefinitely it is not possible to lay down any general rules. But with the molten masses which are injected into fissures and cavities in the earth's crust it is quite otherwise. The degree of liquidity is here also of some importance, but it is subordinate to the form of the space into which the material is driven, or which it has to make for itself by its own kinetic energy. The form in this case will be largely determined by the disposition of the planes of weakness, i.e. of the divisional planes, in the surrounding rocks.

The simplest case of all, perhaps, is when the molten material merely fills a crack in the ground, vertical or otherwise. It may or may not reach the surface, according to circumstances. Such a mass is called a *dyke*. If the crack reaches the surface molten material may spread out as a thin sheet, just as with a lava-flow; in fact one well-known type of volcanic eruption, the fissure-eruption, is exactly of this kind; lava wells up through a crack and spreads out on the surface; the dyke in this case is the feeder of the flow. Dykes are not necessarily vertical at first and may subsequently be tilted at any angle. The name is usually restricted to intrusions which clearly cut across the bedding or foliation planes of the surrounding rocks. When the molten material forces its way as a sheet of varying

Fig. 3. A vertical dyke passing up into a sill which has penetrated along the bedding planes of the sedimentary rocks.

thickness along the bedding planes of a sediment it is called a *sill*. Similarly, more or less horizontal masses cutting across folded and crumpled strata are called *sheets*. The extent and thickness of sills vary indefinitely; some are known to extend over hundreds or even thousands of square miles, e.g. the Whin Sill of the north of England and the Palisade Traps of New York. In some parts of the world, where sills are very abundant, they are the determining features in the topography and hence in the economic value of the land. In the stratified rocks forming the Great Karroo in South Africa, sills are extremely abundant, and by their superior hardness they give rise to the peculiar "kopje" type of scenery, consisting of steep, terraced and often flat-topped hills, with deep valleys between. The kopjes are rocky and barren, but the valleys are often comparatively fertile and well-watered.

Among the large deep-seated intrusions the principal forms are the *laccolith*, the *bysmalith* and the *boss*. The first of these is in general form like a tea-cake, and the strata lying above it are arched up into a dome; at the edges a laccolith often tails

Fig. 4.   A laccolith fed by a dyke below; the overlying strata being lifted up to form a dome.

off into sills and more or less vertical offshoots may form dykes. A bysmalith may be regarded as a laccolith of very great thickness as compared with its horizontal extent, and the overlying strata are of necessity more or less fractured. A boss is a somewhat irregular mass of very great size, which seems to extend indefinitely downwards, such as the Dartmoor granite mass.

In some parts of the world, e.g. Canada, India and South Africa, are found masses of igneous rock, especially of granite, covering hundreds or thousands of square miles, and of unknown depth. Such masses can hardly have been intruded in the ordinary sense of the word; they have more probably been formed by bodily fusion *in situ* of pre-existing rocks while under a very thick cover of later rocks, during a period of depression into a hotter region of the earth.

A fairly common special form of igneous mass is the *neck*. This is in point of fact nothing but the material filling the lower part of the pipe or vent of a volcano, which has

solidified, and has been subsequently exposed at the surface by
the removal of the overlying rocks.   In consequence of their
superior hardness, necks often stand up as elevations above the
softer rocks of the surrounding country; such are many well-
known hills in the central lowlands of Scotland, e.g. North
Berwick Law, Largo Law, Arthur's Seat, etc.

Fig. 5. A volcanic neck which has been denuded and now
stands up to form a hill.

Besides the more or less definite forms mentioned above,
igneous rocks often occur in shapeless masses of every possible
size, their forms depending for the most part on the arrangement
of the planes of weakness in the surrounding rocks.   An extreme
case is where molten material has been injected in numerous
thin sheets along the bedding planes of sediments or the foliation
planes of metamorphic rocks.   This constitutes the so-called
lit-par-lit injection, which is common in many areas of ancient
crystalline rocks, giving rise to banded gneisses.

Divisional planes in rocks.   When any exposure of rocks is
examined, whether natural or artificial, perhaps the first thing
to be noticed is that the masses are not continuous.   In almost
every case the rocks are broken up by planes of discontinuity
into natural blocks of very various shapes and sizes.   These
divisional planes are of several different kinds and originate in
various ways.   The existence and arrangement of divisional
planes in rocks is of immense practical importance, since on
it depends the facility or otherwise of obtaining blocks of stone
of convenient sizes, for various purposes, besides a great
influence on such matters as mining, water-supply, drainage,
etc.

Some divisional planes are original, having been in existence since the first formation of the rock, while others are secondary or superinduced as the result of various disturbing influences. Again some are peculiar to igneous rocks, others to sediments or metamorphic rocks; hence it becomes necessary to treat them systematically and in detail. It may be mentioned here that some divisional planes actually exist as discontinuities in the undisturbed rocks, the *open joints* of the quarryman, while others are rather planes of potential division, along which the rock can be split easily, though in its undisturbed condition it is continuous; an example of this is afforded by ordinary roofing-slates.

The most important divisional planes of rocks are: (1) stratification, lamination or bedding; (2) joints; (3) cleavage, schistosity and foliation; (4) folds; (5) faults. Of these groups the third and fourth are always secondary; the second generally so, while the first group are always original.

Stratification and bedding. Since the sedimentary rocks are formed by accumulations of material for the most part laid down in water, but sometimes on the surface of the land, there is naturally a strong tendency for this material to form horizontal or nearly horizontal layers. Since the nature of the sediment also varies from time to time there is often an alternation of layers of different composition, texture and colour. To express these relations the terms stratification, bedding and lamination are employed. The distinctions between these three terms are not very precise, but generally speaking the word stratification is used to express the succession on a large scale of different kinds of sediment, when exposed to view over a large surface, such as a cliff or a deep quarry; bedding is applied to successive discontinuous layers of sediment of the same kind, while the term lamination is employed for the minute divisional planes, sometimes to the number of hundreds in the space of an inch, which are peculiar to certain fine-textured rocks (see p. 87). However, the employment of these terms by different writers shows much variation, and it is obvious that no sharp distinction can exist between them.

As above stated the layers or beds of sedimentary rocks are usually at first horizontal, but they do not always remain so.

As a result of movements in the crust of the earth, often accompanied by dislocation, the strata are tilted in various ways, making different angles with the horizontal plane; it therefore becomes necessary to have some precise method of expressing this relation. Stratified rocks which are inclined to the horizontal plane are said to *dip*. The dip of a rock-stratum is the steepest line which can be drawn in the plane of its stratification, and the angle of dip is the angle which this line makes with the horizontal, measured in degrees. This fixes the *amount* of the dip, but it is also necessary to find some expression for its *direction*. This is expressed in terms of points of the compass. Thus when a bed is inclined, e.g., to the northeast at an angle of 35°, this may be written shortly as "dip N.E. 35°." If the direction does not exactly coincide with the major points of the compass, it may be expressed as for example "10° east of north," or more shortly "dip N. 10° E." On geological maps the direction and amount of dip is shown by a small arrow, with a figure alongside, indicating the angle of inclination with the horizontal.

The *strike* of a bed is a horizontal line drawn at right angles to the dip. Its direction is also expressed with reference to the points of the compass. For example the full definition of the position of a certain bed might be as follows: "Strike east and west, dip 30° to the north," or more shortly, "strike E.–W. dip 30° N."

It is clear that if a series of uniformly inclined strata come to the surface on level ground, they will intersect the ground in straight lines, which are parallel to the strike; the portion of the rocks then forming the surface is called the *outcrop* of the strata. If however the surface is undulating the form of the outcrop will also be curved and the relations become more complex. This subject cannot here be pursued further, but will be dealt with again in the chapter on geological maps.

Joints. It may generally be observed in any natural or artificial exposure that the rocks are divided into blocks of various sizes and shapes by actual discontinuities, which are sometimes open fissures of varying width. These are called *joints*. They are found both in igneous and in sedimentary rocks, and may come into existence at almost any period of the rock's history. Joints may be conveniently classified into two groups, according to their manner of origin, namely joints due to contraction, and joints due to disturbance.

When a mass of molten igneous rock is cooling and consolidating it undergoes a considerable diminution of volume; since the mass is as a rule unable to contract as a whole, it must of necessity break up into separate portions owing to the strains which are set up in it. If the strains are uniform or nearly so, as is often the case, the blocks thus produced will also tend to be uniform in size and shape. Perhaps the best example of this is afforded by the well-known columnar jointing, so characteristic of certain lavas and intrusive sills. Familiar instances are the Giant's Causeway in Antrim and Fingal's Cave in the island of Staffa, where beds of igneous rock are found to be broken up into aggregates of well-formed columns, for the most part hexagonal in shape, though some have five or seven sides; cross-jointing is also conspicuous, so that the rock tends to break up into blocks rather like cheeses in form. A similar kind of columnar structure is often seen in layers of fine-textured homogeneous sediment, clay or mud, which have been dried by exposure to the air and sun. In all cases the columns are arranged with their long axes perpendicular to the surface of cooling or of drying.

In the case of the larger intrusive masses, such as bosses and laccoliths, the jointing is less regular, and very often the principal joints tend to be arranged parallel to the outer surface of the intrusion; hence if this is curved, so also will the joints be curved. Tabular jointing is also very characteristic of many granite masses, as in Devon and Cornwall.

Joints in stratified rocks are due to two principal causes; contraction on drying, and disturbance during earth-movement. They very commonly tend to arrange themselves in three sets at right angles to one another, breaking up the strata into more or less cubical blocks, which are often of large size. In undisturbed strata two sets of these principal joints or *major joints* are vertical; the third set parallel to the bedding planes. Besides these there are nearly always minor joints of less regular disposition. The presence of many minor joints is a drawback in building stones, as they spoil the shape of the blocks. When stratified rocks are inclined, the major joints no longer remain vertical and horizontal, but, as a little consideration will show, one set is parallel to the dip of the rocks, another to the strike; hence they are called *dip-joints* and *strike-joints* respectively[1].

When strata are tilted from the vertical, and especially when the bedding planes are bent or folded, well-marked systems of joints are often produced. Rocks are not ordinarily plastic, but rather brittle, and they fracture readily owing to disturbance; hence joints which already exist are widened, and new ones are produced. In extreme cases the rocks are much shattered and broken up into numerous, often shapeless blocks of all sizes.

Joints which originate in the ways above mentioned are often widened subsequently by weathering and especially by solution, eventually forming fissures and cavities in the rocks. These are especially prevalent in limestones, which are more soluble than most other common rocks.

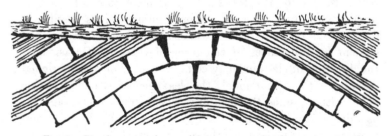

Fig. 6. Development of open joints as a result of curvature in stratified rocks.

Open joints are particularly liable to be formed in well-stratified rocks which have been bent into curved forms, as shown in Fig. 6, since the strains are here unequal and tension is set up; if the bending is sharp, differential movements may take place along the bedding planes, and as explained later, folds, when acute, tend to pass over into actual discontinuities.

---

[1] Most of the problems connected with bedding, dip, strike and jointing in stratified rocks can be well illustrated by means of piles of books.

Cleavage. This is the property possessed by certain rocks of splitting indefinitely into flakes or laminae, which are all bounded by parallel surfaces. The property is best seen in ordinary roofing-slates, which owe their practical value to its presence. Cleavage of rocks is a secondary or superinduced structure, due to pressure, and it therefore belongs properly to the domain of metamorphism. Fine-textured rocks, such as clay or mudstone, consist of vast numbers of minute particles of all shapes, but many of these possess a more or less flattened or elongated form. When first deposited in water they lie in all directions, but under the influence of pressure they are rearranged, usually with their long axes perpendicular to the direction of the pressure. It is evident that a rock possessing a platy structure of this kind will split more readily parallel to the flat faces of the fragments than in a direction transverse to this. The production of cleavage is also assisted by the fact that under pressure a certain amount of mineralogical change usually takes place, leading especially to the formation of mica and other flaky minerals, which split readily. Sometimes the reconstruction of the rocks is so complete that all trace of original structures is lost, but more commonly it is possible to discern the original bedding planes, which are often made evident by differences of colour or of texture. Occasionally the direction of cleavage may coincide with that of the original bedding, but more commonly the cleavage makes a considerable angle with it, since the pressures which produce cleavage are usually horizontal or nearly so in direction and due to compression or crumpling in the earth's crust. Care must be taken to distinguish between cleavage in rocks, which is a secondary structure and variable in direction, and cleavage in minerals, which is one of the original physical properties and is constant in direction in all specimens of the same mineral.

Schistosity and foliation. These terms are applied in a somewhat vague manner to a property analogous to cleavage, developed less regularly in rocks of a coarser texture. The structures produced are generally parallel on a large scale, but less regular and less perfect than in slates. The folia also are often bent and contorted in various ways, so that flat slabs

cannot usually be obtained. The schistose and foliated rocks
have commonly undergone a good deal of mineralogical change
and are often highly crystalline. The term *schist* is generally
applied to those varieties which are rather fine in texture and
rich in mica or other minerals of a somewhat silky or metallic
appearance, while the coarse-grained foliated varieties, which
are rich in quartz and felspar, are called *gneisses*, but no hard
and fast line can be drawn between them. The term foliation
is also sometimes applied to a parallel structure in igneous
rocks, which is not due to pressure, but to the drawing out
during flow of a partly consolidated molten mass of heterogeneous composition and varying colours. This may also be
called *primary gneissic banding*, to distinguish it from the
secondary gneissic banding due to pressure. The schists and
gneisses are typical examples of metamorphic rocks.

Folds. The crust of the earth has been in times past and
still is subjected to strain. The strains mostly take the form
of compression, due to contraction of the crust as a whole.
At and close to the surface this compression generally leads to
fractures, but at greater depths stratified rocks may undergo
crumpling, the originally horizontal bedding planes being
doubled up into arches and troughs of various forms. This
process, which is known as folding, may occur on any scale and
of any degree of intensity. Folds on the largest scale form the
continents and ocean basins; those of an intermediate character
give rise to mountain ranges and other geographical features,
while folds of the smaller kind grade down from this into
sometimes microscopic contortions of bedding planes, such as
are seen in certain schists.

When the strata have been elevated round a central point so
that they dip away from it on every side, this is called a dome,
the converse, where the dip is towards a central point, being a
basin. More commonly however the folding takes place in
elongated areas, the central line of such an area being called
the axis of the fold. Such folds possess a similarity to waves.
A simple elongated arch is called an anticline and the corresponding trough a syncline. Very commonly, a folded area is
made up of a series of parallel arches and troughs, which may

be either symmetrical or asymmetric. Symmetrical folds are formed when the pressure is equal from both sides; if it is much stronger from one side the fold becomes asymmetric, over-turned, or recumbent, according to the degree of asymmetry.

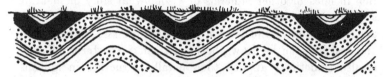

Fig. 7.   Stratified rocks folded into a series of anticlines and synclines.

Fig. 8.   A recumbent fold.

All of these varieties are most easily understood from the figures. Another term commonly employed is isoclinal folding, where both limbs of an overturned fold dip in the same direction.

Folds again may be complex; that is a number of small folds may be combined into the general form of one large fold, as· shown in the figure of an anticlinorium (Fig. 9). An exaggerated form of an anticlinorium is the fan-structure of

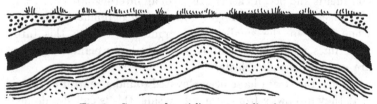

Fig. 9.   Compound anticline or anticlinorium.

some mountain chains, while the enormously long recumbent folds of the Alps may also be mentioned.

Highly-folded rocks commonly form mountain regions, at any rate at first, while those which are less folded constitute regions of lower relief. It has frequently happened however that in course of time elevated areas have been worn down by denudation till they constitute a plain, which may eventually be overflowed by the sea, or by a lake, or covered up by terrestrial deposits. In such a case the newer sediments will rest on the denuded edges of the folded rocks below, producing a strongly marked *unconformity*, and such are very common in the older strata. It is quite clear that in the case of an unconformity the rocks below may be and often are much more folded than those above, while the contrary case is impossible.

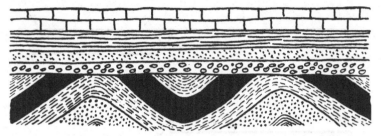

Fig. 10.   Simple unconformity; horizontal strata resting on folded strata.

The importance of the occurrence of rock-folds from the practical point of view is easily demonstrated, especially in mining geology, but it is also of much significance in other ways. Strongly developed folding frequently leads to the outcrop of the same bed at the surface many times within a limited area, and this may have much influence on the topography of the ground, the character of the soil, the water supply and the natural drainage.

Faults. As already explained rocks have often been subjected to strains as a result of earth-movements. Under some conditions these strains result in crumpling and folding of the crust, but in other circumstances, the strata may be fractured instead of folded. Again it sometimes happens that fracture is the result of folding carried beyond the limits of plastic deformation, folds thus passing over into faults. Faults are of many kinds, but the essential feature is differential movement between blocks of the crust, along a plane of discontinuity. Faults may be vertical or inclined at any angle. When vertical, the displacement or distance between the broken ends of the same bed is called the *throw* of the fault. When the fault is inclined, the displacement along the fault-plane may be regarded as made up of two components, the vertical one being called the *throw* and the horizontal one the *heave* of the fault. The inclination of the fault is most conveniently measured by the angle that it makes with the horizontal, or *dip*, as in the case of stratified rocks, although it has till lately been the custom to measure the angle from the vertical; it is then called the *hade* of the fault. All these relations are shown in Fig. 11.

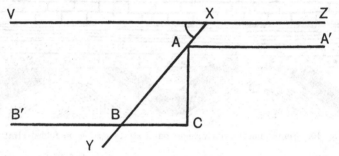

Fig. 11. VZ = horizontal plane, XY = fault-plane. AA', BB' = a horizontal stratum of rock. AB = displacement of fault, AC = throw, BC = heave. $VXY = ABC$ = dip of fault, BAC = hade of fault. BB' is on the downthrow side. AA' on the upthrow side.

It is necessary also to have some means of expressing the relative movement of the two crust-blocks; the one which is moved upwards or away from the earth's centre, relatively to the other is called the upthrow side of the fault, and the other the downthrow side; as a rule it is not possible to determine which block actually moved; it is only the relative displacement that can be measured.

When the fault is vertical and the strata horizontal the relations are very simple, the only quantity required to be determined being the throw or vertical displacement. With inclined strata also there is little difficulty; the effect of faulting may be a repetition of the outcrops of certain strata, as shown in Fig. 12. When both strata and fault are inclined, and in some cases even with vertical faults, the outcrops of some strata may be suppressed altogether, as can be seen from a study of the figures. The possible cases not here figured can easily be worked out by the student.

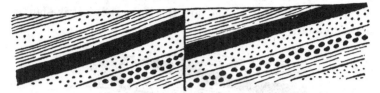

Fig. 12. The outcrop of inclined strata repeated by a fault.

When an inclined fault dips towards the downthrow side (the commonest case) the fault is said to be *normal*. When on the other hand one block has been as it were pushed up the fault-plane this is called a reversed fault. In this case it is always possible by sinking a vertical shaft to pierce some of the strata twice and this may be of importance in mining. On the other hand in the case of a normal fault a shaft may miss a particular stratum altogether (see Fig. 13).

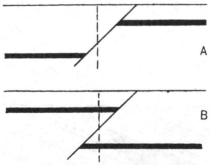

Fig. 13. *A* =normal fault, *B* =reversed fault. The dotted line in each case indicates a vertical shaft or boring cutting the fault-plane.

As already stated a fault may be inclined at any angle; in some instances of reversed faults the inclination approaches the horizontal. Such a fault is commonly called a *thrust-plane*. In extreme cases the displacement may be measured by miles. A thrust-plane commonly results from fracture along the middle limb of a recumbent fold, when the strain exceeds the elastic limit of the rocks.

Since faults frequently result from excessive folding their position often has a definite relation to the direction of the folds; they are generally either parallel or perpendicular to the strike of the rocks, and are therefore called strike-faults and dip-faults.

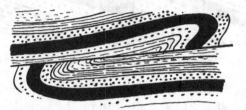

Fig. 14. Recumbent fold passing into thrust-plane.

Each of these naturally produces a characteristic effect on the outcrops; strike-faults cause doubling or suppression of outcrops as before explained (see Figs. 12 and 15). The chief effect of dip-faults is to cause an apparent lateral shifting of the outcrops. The subject is complicated and cannot be pursued here owing to considerations of space.

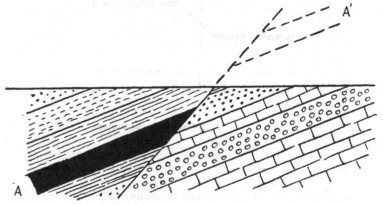

Fig. 15. Suppression of outcrop by faulting. The dotted lines at A' show the original continuation, before denudation, of the bed A, which now does not come to the surface.

Hitherto it has been assumed in the case of vertical faults that the movement was strictly vertical, but this is not always so. There may also be differential movement between the

crust blocks in a horizontal direction, or the displacement may
even be exclusively of this nature.   Here also there will evidently
be lateral shifting of the outcrops in inclined strata.

In nature fault-planes are not as a rule perfectly clean-cut
and straight;   they are often curved or undulating.   During
the movement of the blocks also minor projections are often
broken away, and a certain amount of fracturing and rolling of
material may take place.   Hence fault-fissures are often filled
with fragmental material, known as *fault-breccia*.   Again fault
fissures often serve as channels for the passage of solutions, and
these may deposit material from solution in the fissure.   Hence
faults are frequently filled with secondary material, sometimes
including many minerals of economic value.   Most of the
metallic ore-veins of Cornwall, for example, are found to occupy
fault-fissures, running either parallel or perpendicular to the
strike of the rocks, which is uniform over large areas;   hence
a map of the Cornish mineral veins shows a well-marked
rectangular arrangement.

Conformable and unconformable strata.   The stratified
rocks are laid down in successive horizontal sheets, one above
the other, so long as the conditions remain uniform.   But
rock-formation is a slow process, and during the course of
geological history it has often happened that the crust of the
earth has been disturbed, thus interrupting the regular succession
of strata.   Furthermore, as a result of this disturbance, the
strata may be uplifted and brought within the influence of
agents of denudation, as described in a later section, parts of
them thus being destroyed and the raw edges of the strata
forming the surface of the ground, or sinking again below the
level of the sea.   Subsequently new strata may be laid down
discordantly on these upturned edges, thus giving rise to what
is known as an *unconformity*, which is really a break in the
succession.   There are two chief types of unconformity, known
as *overstep* and *overlap* respectively.

In the first type the upper series has a horizontal base,
resting on the edges of the older upturned rocks, which may be
merely tilted, or folded in a variety of ways (see Fig. 16).   In an
unconformity with overlap the base of the upper series is not

horizontal, but each bed extends further in a given direction than the one below (see Fig. 17). This is the commonest type, where deposition is taking place in the sea. Unconformities are of the greatest possible importance in stratigraphical geology, since they indicate the periods of crust disturbance in the earth's history and provide a convenient means of dividing the stratified rocks into groups of different ages, strongly

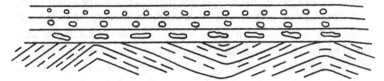

Fig. 16. An unconformity with overstep; the upper horizontal strata rest on the denuded edges of the lower folded strata.

marked unconformities being often used as the boundaries of the rock-systems. They indicate in fact the leading dates in geological history, comparable to the landing of Julius Caesar or the Norman conquest in English history.

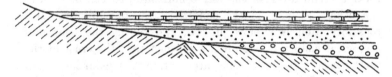

Fig. 17. An unconformity with overlap; each bed of the upper series extends further to the left than the bed below

Textures of rocks. In conformity with modern American usage this term is here employed to designate those intimate features of rocks which are commonly visible only on a close inspection, and are not revealed by a general survey of rock-masses as a whole, or even on a distant view, as is the case with many of the structures just described. The properties known as textures are always, or almost always, inherent in the rocks from the beginning of their existence in the present form. As might be expected the textures of the igneous and of the sedimentary rocks are fundamentally different, and must be described separately. The textures of the sediments are mostly dependent on the absolute and relative sizes of their component particles, these particles having been as a rule brought together by purely mechanical means, and subsequently cemented into a coherent mass by chemical processes. Certain rocks usually assigned to this class are crystalline, having been formed by evaporation of solutions, while others are of organic origin. The textures of all these varieties will be described in later sections, treating of the sediments in detail.

The textures of the igneous rocks on the other hand arise directly from the conditions of their crystallization or consolidation from the fused state. They are therefore controlled by the ordinary physico-chemical laws of solution. If the cooling is very rapid and under low pressure the whole mass may form a perfectly homogeneous solid showing no differentiation into individual crystals. This is known as a *glass*, and is most common among the volcanic rocks. When the cooling is sufficiently slow the fused mass forms an aggregate of crystals of various minerals, the nature of these depending on the original composition of the material. The size of the crystals also is directly determined by the rate of cooling, which in most cases really means the thickness of cover. The size of crystals is practically unlimited, ranging from ultramicroscopic dimensions to large individuals, according to circumstances. Crystalline rocks may therefore be conveniently spoken of as cryptocrystalline, microcrystalline and visibly crystalline, the latter division showing the greatest range of size. Under special conditions the crystals of igneous rocks may be very large indeed, being sometimes measurable by feet; such abnormally coarse-grained varieties are generally called pegmatite, the term referring solely to size of individual crystals, and not to composition. Pegmatites generally occur in the form of veins and dykes, either cutting other igneous rocks or traversing sedimentary or metamorphic rocks.

It is very commonly the case in the igneous rocks that the crystals of one or more minerals are conspicuously larger than the others, owing to crystallization in two stages under different conditions. This is known as porphyritic texture, and is specially characteristic of volcanic rocks, though it is also found

in the other groups. In some cases well-formed crystals which
have been brought up from below are embedded in a glassy
ground-mass formed at the surface in a lava-flow. Other textures specially characteristic of volcanic rocks are those known
as vesicular and amygdaloidal. Vesicles are simply hollows in
the rock, occupied when highly heated by bubbles of steam or
gas, empty when cold. This produces a spongy appearance,
such as is seen in pumice-stone, or in artificial slag from a blast-
furnace. After consolidation of the rock the vesicles are often
infilled by deposits of various minerals brought in by percolating
solutions. The rock is then said to be amygdaloidal, from a
supposed resemblance to almonds in a cake. Volcanic rocks
also often possess a streaky appearance due to flowing and rolling
movements in a viscous heterogeneous mass, and the surfaces,
both upper and under, of lava-flows often show a cindery or
slaggy (scoriaceous) appearance, while the upper surfaces are
often ropy or corded owing to movement in a viscous mass.

The igneous rocks. Although rocks of igneous origin must
be regarded as the original source of all the material composing
sedimentary rocks and soils, they are nevertheless of but
subsidiary importance to the agriculturist, especially in the
British Isles, where they occupy but a small area·of the land-
surface, and this mostly in hilly and little cultivated regions.
In other parts of the world, e.g. Canada, India and South Africa,
rocks of igneous origin cover vast stretches of country, though
they are often masked by surface deposits transported from
afar. Again in many volcanic districts, e.g. Italy, Java and
Central America, the soils are largely formed by weathering
and decomposition of recent lavas. Taking the world as a
whole however there can be no doubt that most of the highly
cultivated regions lie on sedimentary rocks.

An igneous rock may be defined as one which has been
formed directly by consolidation from a state of fusion; an
instance of this may be seen in the solidification of a lava-flow
emitted from a volcano. In this case the cooling takes place
rapidly, under atmospheric pressure only, allowing little time
for the growth of large crystals. Under such conditions the
resulting rock may consist of an aggregate of small crystals of

various minerals, or it may form a completely homogeneous mass without crystalline structure, which is known to petrologists as a *glass*. Such are obsidian, pitchstone and pumice.

On the other hand portions of molten material are frequently injected into cavities formed in the solid crust of the earth at varying depths. When injected under a thick cover of rock, at great depths, the molten material will naturally cool slowly and will also be subjected to high pressure. These conditions preclude the formation of glass and favour the development of large crystals of minerals. An example of a rock formed under such conditions is granite. As would naturally be expected, small masses of rock injected near the surface under a thin cover, cool fairly quickly and therefore partake to some extent of the characters of both groups.

The chemical composition of rocks in general has already been described; it is therefore unnecessary to give here a list of the chemical constituents of the igneous rocks, since all the constituents before enumerated are found in rocks of all classes. The combination of these constituents on cooling and crystallization gives rise to minerals; of the most important of these a list has also been given, and their properties described (see p. 5).

As a result of recent investigations it has been shown that the solidification of molten rock-material takes place according to the laws of solutions; under definite conditions of composition, temperature and pressure, the solution or *magma* will give rise to certain minerals or combinations of minerals. In all cases one of the most important controlling factors is the proportion of silica in the magma, since this governs the composition and characters of the minerals formed. Most of the minerals of the igneous rocks are silicates, i.e. compounds of silica with metals, or in other words metallic salts of silicic acid. Hence silica is regarded as the acid constituent of the magma, all others being bases. A magma rich in silica is called acid, and one poor in silica basic, and the same nomenclature applies to the rocks after solidification, whether crystalline or glassy[1].

---

[1] Harker, *The Natural History of Igneous Rocks*, London, 1909, p. 169.

On the basis of their percentage of silica the igneous rocks can be divided arbitrarily into four groups, as follows:

Silica over 65 per cent.	Acid.
Silica between 65 and 52 per cent.	Intermediate.
Silica between 52 and 45 per cent.	Basic.
Silica below 45 per cent.	Ultrabasic.

These groups, though arbitrary, correspond to real differences of mineralogical constitution. Thus taking as an example the completely crystalline deep-seated or *plutonic* rocks the general characteristics of each group can be tabulated as follows:

Acid group ...	Quartz abundant; alkali-felspar characteristic, and dominant over ferromagnesian silicates.
Intermediate group	Quartz rare or absent; alkali-felspar or plagioclase dominant over ferromagnesian silicates.
Basic group ...	Ferromagnesian silicates dominant over felspar; oxides of iron abundant.
Ultrabasic group ...	Felspar absent; ferromagnesian silicates and oxides the sole constituents.

When the composition of a sufficiently large number of rock-types had been determined it was found that while the acid and ultrabasic groups were simple and well-marked, showing much uniformity, the intermediate and basic groups were each clearly divided into two sub-groups, characterised by excess of potash and soda in one case, and of lime in the other. These are called the alkaline and subalkaline groups respectively and the full scheme of classification may be best expressed as in the table below:

$$\text{ultrabasic} \left\langle \begin{array}{l} \text{basic alkaline} \ -\ \text{intermediate alkaline} \\ \text{basic subalkaline} - \text{intermediate subalkaline} \end{array} \right\rangle \text{acid.}$$

The igneous rocks are thus divided into six groups characterized by chemical differences, to which the mineral constitution of each closely corresponds[1].

---

[1] For a full discussion of the classification of the igneous rocks see Hatch, *Textbook of Petrology*, London, 1914. In this work will also be found detailed descriptions of all the important occurrences of such rocks in the British Isles.

The above diagram may now be conveniently translated into the terms commonly employed by petrologists for these natural groups, taking as an example, as before, the plutonic rocks:

$$\text{peridotite} \left\langle \begin{array}{l} \text{alkali gabbro—syenite} \\ \text{gabbro \quad —diorite} \end{array} \right\rangle \text{granite}.$$

These names are quite arbitrary and for the most part originally meaningless, but they are now accepted as implying a certain limited range of chemical and mineralogical composition.

Name	% silica	Essential minerals
Granite	... Over 65	Quartz, alkali felspar and muscovite, biotite, hornblende or augite.
Syenite	... 65–52	Alkali felspar and biotite, hornblende or augite, with sometimes nepheline, leucite or sodalite.
Diorite	... 65–52	Plagioclase felspar, with biotite, hornblende or augite; often some quartz.
Alkali gabbro	52–45	Alkali felspar (with or without plagioclase, nepheline or leucite) and olivine or augite or both.
Gabbro	... 52–45	Plagioclase felspar, with olivine, augite or hypersthene.
Peridotite	... Below 45	Exclusively composed of olivine, augite, hypersthene, hornblende, etc., with iron ores.

Note. Nepheline, leucite and sodalite are minerals allied to the felspars in composition, but with less silica; they are only found in intermediate and basic alkaline rocks. Hypersthene is closely related to augite.

Exactly similar considerations apply to the volcanic rocks formed from lavas extruded on the earth's surface, except that in many cases the presence of more or less homogeneous glassy matter has to be taken into account. Hence the mineral constitution of the volcanic rocks is less definite. However, when crystalline, they contain the same minerals as the corresponding plutonic group. They may be arranged in a similar diagram:

$$\text{limburgite} \left\langle \begin{array}{l} \text{alkali basalt—trachyte} \\ \quad \text{basalt} \quad \text{—andesite} \end{array} \right\rangle \text{rhyolite.}$$

The rocks of the acid group when completely glassy are generally called obsidian.

Besides the two well-defined groups of the plutonic and volcanic rocks there are as before mentioned a good many occurrences of an intermediate type, rock-masses partaking to some extent of the characters of both. These are for the most part small masses that have been intruded near the surface, under a thin cover, so that cooling was comparatively rapid, though not rapid enough to give rise to the characteristic structures of volcanic rocks. These are called the *hypabyssal* rocks; they occur for the most part as dykes or sills; they are commonly in the main crystalline though many varieties contain more or less glass. This group on the whole conforms also to the classification outlined above, though aberrant forms are rather numerous; the names applied to the normal types are shown in the diagram below:

$$\text{peridotite} \left\langle \begin{array}{l} \text{teschenite— porphyry} \\ \text{dolerite —porphyrite} \end{array} \right\rangle \text{quartz porphyry.}$$

It will be noticed that the name here assigned to the ultra-basic group is the same as that assigned to the corresponding subdivision of the plutonic rocks, the reason being that ultrabasic rocks occur in small masses only and it has not yet been found possible to distinguish the two families satisfactorily.

The igneous rocks as soil-formers. The function of the igneous rocks as soil-formers is twofold; in the first place by direct decomposition and weathering the minerals themselves form soil, which may either remain in place as a sedentary soil, or may be transported elsewhere. Secondly, the igneous rocks yield materials for the formation of sediments, which may be afterwards consolidated into rocks, yielding soil by their weathering.

From a study of the facts detailed in the foregoing paragraphs it is evident that the igneous rocks vary much in chemical and mineralogical composition, hence the character of the soils formed from them must vary also within wide limits. It is of course obvious that as the acid rocks are rich in silica they will yield quartz abundantly, while from the basic rocks this mineral is absent. Furthermore the acid rocks are richer in potash and soda than the basic rocks, but poorer in magnesia, iron and lime. Again the alkali rocks are richer in potash and soda and poorer in lime than the subalkaline rocks of the same silica percentage. Of the common constituents of rocks, potash and lime are of most importance as plant food and most commonly deficient. It follows then that soils formed from granites and syenites will be rich in potash but poor in lime, while soils formed from diorites and gabbros will contain sufficient lime, but will in all probability be deficient in potash. The precise composition of the soil will of course depend to a large extent on the kind of weathering to which the rocks have been subjected, since under certain circumstances soluble compounds may be formed, to be subsequently removed by percolating water (e.g. formation of kaolin, see p. 52). But ordinarily the relations above indicated hold good for potash and for lime. With regard to phosphoric acid, the mineral apatite, the ultimate source of phosphorus in nature, is almost equally abundant in all varieties of igneous rocks and phosphates are rarely deficient in soils derived from them. The other chemical constituents, alumina, soda, magnesia and iron, are generally present in sufficient quantity in all soils, however formed. Soils derived from igneous rocks on the whole tend to be rich in potash and phosphoric acid, although these substances may not always be present in an available form in large quantity.

The actual processes of weathering and chemical decomposition by which the minerals of the igneous and other rocks are made available in the soil form a subject of the greatest importance from the agricultural point of view. They are treated in detail in Chapter II.

The crystalline schists. These form a large and varied, but distinctive, series of rocks belonging to the metamorphic group. Their origin is in some instances still obscure, but the majority of them have certainly been formed by the alteration

of normal igneous and sedimentary rocks. They generally
show in a high degree the properties of cleavage, schistosity
and foliation, as before defined. Many of them also contain
certain peculiar and distinctive minerals, such as are known to
be developed under conditions of high temperature and high
pressure. Some of these are also common to the igneous rocks,
while others are mainly confined to this group. Most of these
rocks come under the general designation of *gneiss* and *schist*,
but massive and non-foliated rocks, such as quartzite and marble
are also common. The true crystalline schists belong to the
earlier stages of the earth's history and in most parts of the world
they form the foundation on which the later rocks rest. In
fact it is still uncertain whether true crystalline schists have
ever been produced since the very earliest stages of the earth's
history.

Many of the gneisses are to all intents and purposes quite
similar to the igneous rocks, especially those of the granite
group. The parallel structures so often seen in them are often
due to the drawing out during flow of an imperfectly mixed
liquid mass in process of crystallization. This gives a streaky
appearance, with bands, often twisted and contorted, of varying
composition and colour. In other instances a similar appearance
is produced by the recrystallization, under intense pressure, of
the minerals of rocks already formed. During this process
much heat is undoubtedly generated by friction, and this
assists in the formation of new minerals, the constituents of the
rocks combining in a different way. The effects of the high
pressure and temperature is to a large extent to reverse the
effects of weathering, and sediments that have been highly
metamorphosed often take on characters very like those of the
igneous rocks from which they were originally derived.

Besides the minerals found in the igneous rocks, there exist
a whole series of minerals specially characteristic of the
crystalline schists and of the metamorphic rocks in general.
The most important of these are various silicates specially rich
in alumina; such are garnet, andalusite, kyanite, sillimanite,
cordierite and staurolite. In marbles formed from impure
limestones and dolomites there are found also special minerals

containing calcium and magnesium, such as wollastonite, forsterite, tremolite and spinel. None of these are of much importance as soil-formers, and it is unnecessary to give any further account of them. Marbles are of importance only in so far as they yield lime by their decomposition. The other plant-food constituents are as a rule not present in any appreciable quantity. The weathering of the true gneisses follows the same lines as that of the igneous rocks; granitic gneisses yield soils rich in potash and phosphoric acid, while nitrogen is always deficient. The schistose rocks formed from the clay-like sediments form generally heavy soils, of very variable character, and the final result of their weathering depends for the most part on the climatic conditions, and especially on the amount and seasonal distribution of the rainfall. If this is heavy, most of the soluble constituents are washed out, and the result may be a poor, hungry soil, whereas in a dry climate plant-food may be concentrated and irrigation will yield heavy crops.

CHAPTER II

WEATHERING

Introduction. According to the accepted definition the soil is to be regarded as the uppermost weathered and disintegrated layer of the earth's crust; its material is derived either directly or indirectly from rocks, together with a certain amount from the atmosphere. Any investigation of the origin and characters of the soil accordingly involves a knowledge of the processes of disintegration and of chemical and mineralogical change which take place in the constituents of the rocks. These changes may be summed up in the general term *weathering*, since they are to a large extent due to atmospheric or meteoric agencies. The vital activities of animals and plants also play an important part.

Most of the processes of weathering are of a complex nature, being frequently the result of several causes acting simultaneously, and it is often difficult to ascertain clearly how much of the final result is to be attributed to each agent concerned. Hence a certain amount of repetition is almost unavoidable. A considerable degree of obscurity also prevails in some parts of the subject, and widely divergent views still continue to be held as to the exact nature of the processes by which certain well-known and definite results are obtained; as an example may be mentioned the decomposition of the felspars, and of other common silicates. Much work is still required in this field of investigation.

Kinds of weathering. As a matter of convenience weathering may be classified under three general headings, namely, physical, chemical and organic. Under physical weathering are included

all those processes resulting in a change in the state of aggregation of the mineral particles of a rock, unaccompanied by any
chemical or mineralogical alteration. The rocks are thus
broken up into smaller portions or even into their constituent
mineral grains. This in itself promotes the state of fine division
which is necessary in a soil, and by increasing the free surface
area of the particles it also facilitates the action of chemical
weathering agents. The second group includes all those changes
resulting in an alteration of the composition, not only of the
rock as a whole, but also of individual particles of it. This
necessarily involves mineralogical changes which are often far-
reaching in their character. One of the most important results
of chemical weathering is, in a general way, to render the
supplies of plant-food contained in the rocks more readily
available for assimilation by the plant, thus increasing the
fertility of the soil. Chemical weathering also produces
exceedingly important results by leading directly to disintegration and subdivision of the soil particles, often accompanied
by removal of certain constituents in solution or in suspension.
The effects of the vital activity of animals and plants are partly
physical, assisting disintegration, and partly of a chemical
nature. The latter class of changes are often highly complex,
involving the formation of various chemical compounds whose
exact character is still obscure. In particular, the action of
bacteria and other protozoa is undoubtedly of great importance
in soil formation, but on this subject the information at our
disposal is scanty and somewhat contradictory. It is well
known that bacteria play an important part in certain processes
in the soil, especially in connexion with the formation of
nitrates, but it is highly probable that they have an important
influence also in many other ways, which still remain to be
worked out in detail[1].

Climate and weathering. Among the most important
results of recent investigations into the origin and characters of
soils, carried out for the most part in Germany and Russia, has
been the recognition of the great influence of climatic conditions
in determining the ultimate character of the soil. Since

[1] Russell, *The Fertility of the Soil*, Cambridge, 1913, Chapters I and

physical and chemical processes are controlled to a great extent
by temperature and the presence or absence of water, and since
these are also important factors in climate, it follows that the
type of weathering in any given area must also be controlled
by these factors.

Owing to meteorological causes, into which it is unnecessary
to enter here, the whole earth can be divided into seven
climatic zones or belts, as follows:

The *equatorial* or *tropical zone,* on either side of the equator,
characterized by high temperature and heavy rainfall, leading
to luxuriant vegetation, which tends to facilitate decomposition
of the rocks by chemical and organic agencies; owing to the
protective effect of thick vegetation transport is in abeyance,
and the soils are deep and rich in plant-food. The action of
bacteria, being favoured by the climate, is also at a maximum.

The *desert zones.* On either side of the equatorial belt in
each hemisphere is a region of varying width characterized by
high temperature and very small precipitation, forming the
great rainless regions of the world. Here water action is in
abeyance, while animals and plants are almost completely
absent.

The *temperate zones* are regions of moderate temperature
and as a rule of fairly abundant rainfall, though in this respect
much local variation exists. The greater part of the land-
surface is covered by vegetation and animals are numerous.
The types of weathering which prevail in these zones are of a
mixed character and all the ordinary geological agents are
operative to a greater or less extent. The seasonal variations
of climate are strongly marked, and at different times of the
year different agents of weathering are dominant. The water-
systems of the land are well developed, and transport of material
by running water plays an important part in soil-formation.
Weathering and transport are as a rule nicely balanced and
weathered material does not commonly accumulate in place
to great depths, as in tropical regions.

The *arctic zones* are characterized by extremely low temperature, animal and vegetable life is scanty, and soil-formation is
at a minimum. The most important geological agent is frost,

and the low temperature is unfavourable to chemical weathering. Large surfaces are covered by permanent ice and snow, and bare rock is abundant. The absence of light for part of the year is unfavourable to plant growth, and in truly arctic regions agriculture is non-existent. The actual precipitation is generally small and the low temperature keeps most of the water in the solid form; hence for practical purposes the climate is dry.

The climatic conditions of these zones can be conveniently summarized in a tabular form, as follows:

Tropical zone	...	Hot and damp.
Desert zones	...	Hot and dry.
Temperate zones	...	Cool and damp.
Arctic zones...	...	Cold and dry.

These climatic zones are well marked and on the whole they follow the parallels of latitude, with variations induced by the relative positions of land and sea. An important factor in the latter consideration is the general north-and-south trend of the continents and oceans. Stated in general terms the central parts of the continents tend to be drier than the mean, with greater variations of temperature, i.e. hotter summers and colder winters, while over the oceans the conditions are more equable. Prevailing winds and ocean currents also have an important influence. For example, Western Europe is much warmer in winter than corresponding latitudes in Eastern America or Eastern Asia. Cultivation can consequently be carried on in Norway within the Arctic circle, while Labrador, in the latitude of Britain, is a barren waste. Again the southward-flowing Mozambique current makes the climate of Natal almost tropical. Owing to deficiency of rainfall the interior of Asia is largely desert, and indeed in this region the temperate zone is practically absent; the desert region extending northwards till it meets the frozen swamps of Eastern Siberia[1].

Of almost equal importance with latitude as a climatic influence is height above sea-level. In the mountain ranges of the temperate zone and in the highest mountains of the

[1] For a good account of the general distribution of types of climate see Lake, *Physical Geography*, Cambridge, 1915.

tropics the climatic and geological conditions resemble those of the arctic zones; glaciers abound in many mountains in low latitudes, e.g. the Alps, the Himalaya, the mountains of east-central Africa, the Andes, the mountains of New Zealand and others. In fact everywhere above the permanent snow-line, at whatever height that may happen to be, the conditions are arctic. The presence or absence of water is also of great importance, as witness the fertility of the oases of the Sahara in the desert zone, and of certain arid regions in South Africa and Australia, when artificially irrigated.

Weathering by chemical processes. Under this heading are included all those processes in rocks that lead to disintegration and alteration by chemical means. The chemical processes which operate in this way are numerous and often complex in their action. The following list comprises the most important of them:

1. Solution.
2. Hydration.
3. Hydrolysis.
4. Pneumatolysis.
5. Oxidation.
6. Reduction.
7. Carbonation.

Nearly all the chemical processes which take place in minerals and rocks can be ranged under one or other of these heads; it must however always be remembered that two or more of these processes may be, and generally are, operative at the same time, and it is often a matter of great difficulty to ascertain to which class or classes a given case should be assigned. Thus for example solution, hydration and hydrolysis are very closely allied, while oxidation and nearly all ordinary chemical reactions are greatly facilitated by the presence of water. A familiar example is the rusting of iron; indeed many chemical changes cannot take place at all if the substances are perfectly dry. Again, no natural waters are absolutely pure, and the gases and solids dissolved in the water play an important part in chemical processes of weathering.

Solution. The simplest of all weathering processes is solution. Most minerals are soluble to a certain extent in pure water, though the degree of solubility is in general very small. Some of the salts of sodium, potassium and magnesium, especially the chlorides and sulphates, are however readily soluble. If beds containing these salts are leached out by percolating water, a general subsidence and collapse of overlying strata may take place, possibly forming lake-basins. With the exception of rock-salt, these minerals are very rare in large masses. However they certainly exist largely in a finely divided state in the soils of many regions and their removal by solution has an important bearing on the fertility of the soil. Similar considerations apply in a less degree to calcium sulphate, which exists in two natural forms, called anhydrite and gypsum respectively.

The different forms of calcium carbonate and the mineral dolomite are also comparatively soluble in water and their removal in solution has very important effects on the denudation of limestone regions. This subject will be treated later under the heading of carbonation. Some of the compounds of iron are easily soluble in water, e.g. the chlorides and sulphates, and if present in rocks or soils they are liable to be removed. The solubility of silicate minerals is universally very small and in fact the effect of pure water on most of them is negligible at the atmospheric temperature and pressure. But since the solubility of nearly all minerals is increased by rise of temperature and pressure it follows that at great depths within the earth solution must be more active. This is shown by the fact that hot springs coming from great depths contain much mineral matter derived from the rocks through which the water has passed (see pp. 46 and 52).

Hydration. This process is of very common occurrence in nature and has an important influence on soil-formation. Most of the chemical elements have the power of forming hydroxides in the presence of water, and some do so with great readiness. A familiar example of the process is the *slaking* of quicklime, which is represented by the equation:

$$CaO + H_2O = Ca(OH)_2.$$

The slaking of lime is accompanied by evolution of heat and disintegration or crumbling of the lime. Processes very similar to this, but much less rapid in their action, occur in the natural disintegration of rocks, and lead to important results. Oxides of iron are specially liable to hydration; thus haematite, Fe_2O_3, is hydrated to form limonite, $2Fe_2O_3 . 3H_2O$, and hydrates of iron with different proportions of water exist as natural minerals. Hydrates of alumina, such as bauxite and diaspore, also commonly occur in weathered rocks.

Many minerals also are known consisting of basic salts of certain metals, while others possess water of crystallization. A good example of the latter is the mineral gypsum, which has the formula $CaSO_4 . 2H_2O$, while calcium sulphate also exists naturally as the anhydrous mineral anhydrite, $CaSO_4$. The formation of gypsum from anhydrite, a common process, results in an increase of volume by about 33 per cent., so that it has a powerful disintegrating effect.

Hydrated silicates are also exceedingly common, and their constitution is generally very complex. Such are the great group of the zeolite minerals which are supposed by some authorities to exist largely in soils, though this is still somewhat doubtful. It is highly probable that the formation of many hydrated silicates is really due to hydrolysis, as will be explained in the next section. The formation of chlorite from biotite may perhaps be a case of simple hydration; it is very common in the weathering of granites; it is possible however that there is here in addition some removal of potash in solution.

Hydrolysis. When a salt of a strong base and weak acid, for example, sodium carbonate, is dissolved in water, dissociation occurs and the solution possesses an alkaline reaction. This is due to a decomposition which may be represented by the following equation:

$$Na_2CO_3 + 2H_2O = 2NaOH + H_2CO_3.$$

The alkali felspars may also be regarded as compounds of strong bases, K_2O and Na_2O, with a weak acid, aluminosilicic acid, and in contact with water a similar reaction also takes place, though with extreme slowness. In the case of orthoclase this results in the formation of kaolinite or its amorphous equivalent, halloysite; the change may be represented by the equation:

$$K_2O.Al_2O_3.6SiO_2 + 3H_2O = 2KOH + Al_2O_3.2SiO_2.2H_2O + 4SiO_2.$$

The colloidal silica thus set free is soluble in caustic potash and is removed in solution, thus leaving a kaolinitic residue, which is an important constituent of many soils and in its purest form constitutes china-clay.

Similar reactions occur when any of the common rock-forming silicates of magnesia, lime, etc., are acted on by water, resulting in the formation of hydrated and free silica, together with, in some cases, hydrated silicates of these bases, which appear to be often of rather complex composition. Thus for example the formation of serpentine from olivine is probably to be attributed to hydrolysis, serpentine being a hydrated silicate of magnesia.

Pneumatolysis. This term is used to signify the action of highly heated and therefore chemically active gases on rocks and minerals. Its practical effect in weathering and soil-formation is local and limited, being manifested chiefly in volcanic regions and districts where hot springs are prevalent. The pneumatolytic decomposition of minerals is due partly to highly heated water-vapour and partly to such volatile elements as boron and fluorine. It is specially prevalent in connexion with granite intrusions, this rock being often weathered and shattered down to great depths. A good example is afforded by the formation of kaolinite from the granites of Devon and Cornwall, Karlsbad, etc. This process is essentially the same as hydrolysis, but carried on at a higher temperature and pressure and therefore more active. Perhaps the commonest mineral product of pneumatolysis in granites is tourmaline, which is very stable, and in small quantities a common constituent of sands and soils. The pneumatolytic action of steam and other vapours doubtless accounts in part for the rapid weathering and disintegration of lava-flows, which in a comparatively short time often yield soils of extraordinary fertility. In regions of expiring volcanic activity the ground

is often at a high temperature at a small depth from the surface,
as well as being largely penetrated by heated waters; in such
places decomposition of rocks and minerals is rapid and far-
reaching, most of the common silicate minerals being here
almost completely decomposed, and the rocks much disintegrated. In such regions hot springs often bring to the
surface much dissolved matter, especially carbonate of lime,
silica and various salts of potassium, sodium and magnesium
(mineral springs).

Oxidation. This is undoubtedly a most important process
in weathering and soil-formation, but its effects are always so
closely associated with hydration and other chemical processes
that it is very difficult to give any clear and connected account
of them. Oxidation is most conspicuous in the case of iron
compounds, since the change from the ferrous to the ferric
state generally involves a notable alteration of colour. Ferrous
compounds are generally black, green or grey, whereas ferric
compounds usually show some shade of yellow, brown or red.
This accounts for the commonly observed fact that the soils
overlying dark-coloured rocks, such as basalt, slate or grey clay,
have a colour quite different from that of the unaltered rock,
and in the same way rocks largely composed of black or green
silicates containing iron always have a brown weathered crust.
It can often be observed in deep mines that this change of colour
has extended down to depths measurable by hundreds of feet,
in fact, as far down as the rocks are saturated with water
coming from the surface, carrying with it oxidizing agents, and
especially atmospheric oxygen. Although oxidation is most
readily observable in iron compounds, it also occurs in many
other cases.

Oxidation always involves a change of volume in the minerals
affected and thus produces mechanical stresses, leading to
disintegration. Many highly oxidized substances, such as ferric
compounds, crystallize with difficulty under normal conditions
and tend to remain in a granular and amorphous form, thus
facilitating the formation of a fine tilth.

Oxidation also plays a most important part in the decomposition and decay of organic matter, the ultimate products

being carbon dioxide and water; these processes are most commonly brought about by means of bacteria and are highly complex in their nature. The complex nitrogenous compounds occurring in animal and vegetable tissues are converted by a series of bacterial changes into nitric acid and nitrates, which are of immense importance as sources of plant food. The nature of the organic constituents of the soil will be considered again in a later section.

Reduction. This process, which is the converse of oxidation, is not of much significance in weathering and soil-formation, since the prevailing conditions are generally unfavourable to it. Reduction does take place however in sour water-logged soils, where the oxygen of the air is excluded and anaerobic bacteria flourish. It is on the other hand the prevailing reaction in the formation of sedimentary deposits in deep water and especially in the sea, where ferric compounds are reduced to the ferrous state, resulting in the formation of iron sulphides, either the black sulphide, FeS, or pyrites and marcasite, FeS_2. To the presence of these substances is largely due the generally prevailing blue or grey colour of unweathered clays. This reduction is chiefly brought about by the decomposition of organic matter buried in the sediment, the remains of animals and plants that lived in the sea or were carried down from the land. It is from this decay also that the sulphur is derived for the formation of the sulphides.

Carbonation. This term is generally employed to designate the effect on rocks and minerals of carbon dioxide dissolved in water. This is undoubtedly a most important agent in weathering and denudation, and until recently almost all the chemical alterations in minerals were attributed to it. Of late years however, the idea has begun to gain ground that some of these changes can be explained in other ways, e.g., by hydrolysis (see p. 44).

Carbon dioxide when dissolved in water forms an acid, carbonic acid, which has however never been isolated in the pure state. The solubility of carbon dioxide is increased by pressure, hence natural waters at great depths must have a more powerful effect than those at and near the surface. Under the most favourable circumstances however carbonic acid is a weak acid, and its effects in displacing other acid radicles only become noticeable when continued for a considerable time.

The most conspicuous and easily observed effect of dissolved carbon dioxide is its action on calcium carbonate. When these substances are brought in contact a special reaction occurs and a bicarbonate is formed, thus:

$$CaCO_3 + CO_2 + H_2O = CaH_2(CO_3)_2.$$

This compound is more soluble than the normal carbonate, and it is in this form that calcium carbonate really exists in natural waters. It is unstable and is broken up by heat, or release of pressure, again forming normal carbonate and carbon dioxide. For this reason calcareous waters from deep-seated springs often deposit much carbonate of lime when they reach the surface.

The solvent action of carbonated water plays a very important part in the denudation of limestone regions, and in the weathering and disintegration of rocks containing carbonates, either as their principal constituents or as cements, e.g. calcareous sandstones. In the latter case the cement is removed and the grains of quartz and other minerals are left free and may form sandy residual deposits. In a similar way magnesium and iron compounds may be removed in solution as carbonates, being somewhat soluble in that form, while the carbonates of sodium and potassium are readily soluble salts; these latter are found in considerable quantity in the waters of certain salt lakes, having been carried into them by streams; hence it follows that they must be present in river waters and in natural underground waters generally.

Weathering by physical processes. The agents concerned in physical weathering are almost entirely atmospheric in their origin, consequently they are for the most part strictly dependent on climate. The dominant factors are variations of temperature and presence or absence of moisture. In the first place it is obvious that many of the chemical changes above described must also have an important mechanical effect. Thus solution of the binding material of a cemented rock causes disintegration of the rock as a whole, the insoluble particles being set free. This action alone is of itself sufficient in some cases to form a residual soil. Hydration and oxidation also bring about disintegration by change of volume, and other instances might also be cited. These however are but subsidiary effects. Among purely mechanical agencies of weathering the most important are expansion and contraction consequent on changes of temperature, and the expansion of water in freezing.

It is well known that all natural substances expand and contract in volume as the temperature changes. In nearly every case expansion is caused by rise of temperature, although the absolute amount, or coefficient of expansion, varies in different substances. Again in crystalline minerals change of volume is a vectorial property, having different values in different directions. Since rocks are not as a rule homogeneous, but consist of an aggregate of different minerals, the expansion is usually unequal and differential strains are set up, leading to fracture and shattering of the rock as a whole, or of the individual crystals. It is evident that the more sudden the change of temperature, the greater will be the strains, and the greater the consequent shattering.

The conditions most favourable to a full development of this process are to be found in arid regions in low latitudes, in other words, in the desert zones. Here the temperature by day is extremely high, while after sunset owing to the clear sky radiation is strong and the fall of temperature is very sudden. In consequence of this change the rocks are cracked and shattered to a great extent. From vertical faces and on steep slopes the loosened fragments fall down and form great screes at the foot, but on level surfaces the effect is limited. Rocks are bad conductors and a coating of loosened fragments of comparatively small depth will prevent heating and cooling of the underlying solid rock. Hence in hot, arid regions the hills will be worn away much quicker than the plains, tending towards a general levelling of the surface. The actual daily variations of temperature in such regions are often very great, amounting sometimes to 70° F., and blocks weighing up to several hundred pounds are often broken off with a loud report. But more commonly quite thin scaly layers break off from the surface, and such scaly weathering, the *desquamation* of Richthofen[1], is specially liable to occur in granitic rocks. In temperate climates, such as the British Isles, this kind of physical weathering is probably of minor importance, but in certain parts of the United States, in Texas and California, and even in Massachusetts, it is very prominent[2].

In temperate and cold climates the most important physical agent of disintegration is the expansion of water on freezing. When the rainfall is abundant the crevices and pores of rocks are filled with water and this on freezing expands and exerts an almost irresistible force. The pressure exerted in this way is more than sufficient to overcome the cohesion of all ordinary rocks. The expansion is very considerable; 1 c.c. of water at 0° C. when frozen yields 1·0907 c.c. of ice; the increase of volume is therefore about $\frac{1}{11}$ of the whole. When the temperature falls below 0° C. the water in the rocks freezes, expands and shatters them, but it is only after a thaw that the effects become manifest, since while the water is frozen it welds the whole into a coherent mass, which falls to pieces when the ice melts. The process is well illustrated by the spongy state of an ordinary road or gravel path after a thaw. When the water freezes the whole surface rises bodily, but is hard so long as it is frozen; only when the ice disappears is the increased distance apart of the stones perceptible.

In cold-temperate and arctic regions the shattering effects of frost play a similar part to changes of temperature in the tropics. Blocks of rock of all sizes are broken off, forming screes among mountains, and on level ground covering the surface with a layer of disintegrated rock-debris. This action is of great importance in soil-formation, and on it depend the beneficial effects of autumn and winter ploughing, especially on heavy clay land. The blocks of soil turned up by the plough in the autumn are shattered by the frosts of winter, thus

[1] Richthofen, *Führer für Forschungsreisende*, Berlin, 1886, p. 94. Walther, *Das Gesetz der Wüstenbildung*, Berlin, 1900.
[2] Merrill, *Rocks, Rock-weathering and Soils*, New York, 1897, p. 181.

facilitating the formation of a fine tilth at spring seed-time. The thorough aeration thus brought about also assists to render available the plant food in the soil, by means of oxidation, carbonation and the activity of bacteria. In high mountain regions of the temperate zone and at all elevations in the arctic regions this is by far the most important agent of disintegration, since at low temperatures many chemical processes are in abeyance, and micro-organisms are unable to exist. Most chemical weathering depends directly or indirectly on the presence of water, and when the water is frozen the weathering processes cannot go on.

Although wind can scarcely be considered as a weathering agent in the ordinary sense, yet it does possess a certain power of disintegration through the kinetic energy of transported particles. This may be compared to the sand-blast as employed for etching glass and other hard substances. This natural sand-blast, when long continued, must have some effect in producing very finely comminuted material, and in fact the fine dust of the deserts is well known; it is probably produced largely in this way, partly by friction of loose particles against each other and partly by their grinding effect on solid rocks. Soft minerals and those possessing a good cleavage are generally absent from wind-blown sands, having been reduced to impalpable particles by long-continued attrition.

The weathering of some common rock-forming minerals. The operation of the ordinary processes of weathering can perhaps be most satisfactorily illustrated by a consideration of the chemical and mineralogical changes taking place in the case of some of the commoner minerals constituting the greater part of the rocks of the earth's crust. In nearly all cases these changes are the resultant of several processes acting at once. It is comparatively seldom that they are perfectly simple in character and referable to the influence of one agent alone.

Quartz is one of the most stable of all minerals; at ordinary temperatures and pressures it is unaffected by any of the ordinary chemical reagents, and is only effectively attacked by hydrofluoric acid. Consequently it undergoes no chemical changes, and therefore tends to be concentrated in both residual and transported deposits. The only change which in practice takes place in particles of quartz is a gradual reduction in size, accompanied by rounding, arising from mutual attrition and occasional contact with still harder mineral particles. Hence quartz is without doubt the commonest of all minerals in sedimentary rocks and in soils. It is the principal constituent of sands, and is also abundant in muds and other fine-grained clastic deposits. Grains of quartz may pass on from one stratified deposit to another with little or no change.

The cryptocrystalline and non-crystalline forms of silica are more easily dissolved than quartz, colloidal silica being fairly soluble in alkaline solutions; colloidal silica is a by-product of the decomposition of many silicates, as will appear in detail hereafter.

It has however been shown experimentally by Pfaff and others that under increased pressures and at high temperatures even quartz is appreciably dissolved by water, though but to a small extent. It is clear that the highly superheated water occurring at deep levels, and especially the so-called juvenile waters (i.e. water of volcanic origin) must have the power of dissolving quartz to some extent. Hot springs in volcanic regions (Iceland, the Yellowstone Park, and New Zealand) often contain much dissolved silica, which is deposited as *sinter* around the outflow of the spring, and this may be in part derived from direct solution of quartz at great depths, though much of it probably comes from hydrolytic decomposition of silicates, especially felspars.

The *felspar* group includes a large number of minerals of varying composition and varying degrees of stability, hence they are affected by weathering in various ways. From the practical point of view the most important of these changes is the formation of kaolinite from orthoclase or albite, a process which results in the production of great beds of *china-clay*. The composition of orthoclase is most conveniently represented by the formula $K_2O . Al_2O_3 . 6SiO_2$, while kaolinite is a hydrated silicate of alumina represented by the formula $Al_2O_3 . 2SiO_2 . 2H_2O$.

The reaction by which kaolinite is formed has been very

generally attributed to the influence of carbon dioxide dissolved in water, but according to the more modern view it is to be explained as due to water alone; it is in fact to be regarded as a case of *hydrolysis*. The change can then be represented by the following equation:

$$K_2O . Al_2O_3 . 6SiO_2 + 3H_2O$$
$$= 2KOH + 4SiO_2 + Al_2O_3 . 2SiO_2 . 2H_2O,$$

the resultant products being potassium hydroxide, silica and kaolinite. Now silica in the non-crystalline form is readily soluble in caustic potash, or to state the case more correctly, potash and silica combine to form potassium silicate, which is soluble in water, and is removed, leaving only kaolinite in the solid form. Sometimes this mineral is found crystalline, while in other cases it occurs as the amorphous modification, halloysite, which has the same composition. The crystallization or otherwise of this mineral is probably determined by the temperature at which the change takes place, as crystals are more common in cases where there are indications of the pneumatolytic action of steam at a high temperature.

This hydrolytic decomposition takes place more or less whenever water is in contact with felspar, but in most cases with extreme slowness. It is only when the temperature is high, owing to volcanic, solfataric or hydrothermal action near the surface or to pneumatolysis at greater depths, that it takes place at all rapidly. Under favourable conditions however it may result in the formation of great masses of china-clay, as in Devon and Cornwall, Saxony and near Limoges.

Another and perhaps even commoner type of decomposition affecting orthoclase or microcline is conversion into an aggregate of minute flakes of white mica. This change involves the removal of a part only of the potash together with two thirds of the silica. The composition of typical muscovite is approximately $2H_2O . K_2O . 3Al_2O_3 . 6SiO_2,$

$$3[K_2O . Al_2O_3 . 6SiO_2] + 2H_2O$$
$$= 2K_2O + 2H_2O . K_2O . 3Al_2O_3 . 6SiO_2 + 12SiO_2.$$

The resulting products at ordinary temperatures are therefore muscovite and colloid silica, but at high temperatures the silica may crystallize as quartz, forming a quartz-mica rock, or *greisen*. The former case is the more common and this only is of importance in normal weathering and soil-formation. A precisely similar reaction occurs in the case of soda-felspar, yielding a soda-mica, paragonite, which in minute flakes is indistinguishable from muscovite. This decomposition of alkali-felspars furnishes much of the finely-divided micaceous substance which is so commonly a constituent of soils, of modern sediments and of the older rocks formed from similar sediments[1].

The plagioclase felspars are mixed crystals of two components, albite and anorthite, in varying proportions. The albite molecule is $Na_2O . Al_2O_3 . 6SiO_2$, while the anorthite molecule is $CaO . Al_2O_3 . 2SiO_2$. These two molecules are differently affected by weathering agents, and give rise to different products. This subject is complex and not well understood; the albite molecule appears to form kaolinite or mica, according to circumstances, as in the case of orthoclase, while the lime-felspar molecule gives rise to a variety of products, such as epidote, chlorite and even calcite.

Under the influence of certain peculiar types of weathering, whose nature is even yet not properly understood, the felspars give rise, not to silicates, as previously described, but to various hydroxides of aluminium, the most important being hydrargillite or gibbsite, $Al_2O_3 . 3H_2O$, and bauxite, $Al_2O_3 . 2H_2O$. This type of weathering is specially characteristic of the laterite deposits of tropical regions[2], and appears therefore in all probability to be determined by climatic conditions, or possibly even by the agency of bacteria[3]. (For a discussion of the origin of laterite see pp. 57 and 115.)

It has often been observed that the decomposition products of lime-bearing plagioclase include some newly formed albite, this molecule having apparently recrystallized as such, while the lime-molecule has formed compounds belonging to other

[1] Hutchings, *Geol. Mag.* 1894, pp. 36 and 64.
[2] Bauer, "Beiträge zur Geologie der Seychellen-Inseln, und besonders zur Kenntniss der Laterite," *Neues Jahrb. für Min.* 1898, p. 168.
[3] Holland, "Constitution of Laterite," *Geol. Mag.* 1903, p. 59.

mineral groups. This process is closely analogous to the albitization of igneous rocks of basic composition[1].

The decomposition of lime-bearing plagioclase felspar is perhaps the most important of all sources of lime compounds in the stratified and other sedimentary deposits. To this we must look for the primal origin of the great masses of calcium carbonate which constitute the calcareous rocks, limestone, dolomite, calc-sinter and travertine, as well as the thick beds of gypsum and anhydrite and the lime-content of all fresh waters and of the sea.

The mica group. It is remarkable that a great difference of behaviour towards weathering agents is found in the members of the mica group. Stated in the most general terms, the colourless or very pale alkali-micas of the muscovite subgroup are particularly stable minerals, being very little affected by weathering and consequently persisting through many geological cycles; their final disappearance is due to their very perfect cleavage, which causes them to be easily abraded, rather than to chemical change. Indeed alkali-micas are a very common product of the chemical alteration of silicate minerals, felspars, nepheline, cordierite, garnet, and many others. On the other hand the iron and magnesian micas of the biotite group are very unstable, being easily decomposed to chlorite, epidote and other hydrous silicates. Consequently they are not common constituents of sediments or of soils. The easy decomposition of biotite sets free iron and magnesia, and what is of more importance agriculturally, a considerable amount of potash, which exists in the soil in a form readily available as plant food, probably as solutions of potassium salts absorbed by the argillaceous or organic constituents of the soil.

The amphibole and pyroxene groups. As a matter of convenience and to save repetition the minerals of these two groups may be treated together, since they are very similar in chemical composition and in mineralogical characters, yielding eventually the same weathering products. Essentially they are all mixed silicates of magnesia, iron and lime, often with soda

---

[1] Bailey and Grabham, "Albitization of the Plagioclase Felspars," *Geol. Mag.* 1909, p. 250 (with further references).

and alumina. The amphiboles and pyroxenes are polymorphous forms of similar compounds and can be converted one into the other under varying conditions of temperature and pressure. Thus a very common change in igneous rocks is the conversion of augite into hornblende (uralitization). By weathering both amphiboles and pyroxenes are converted into various silicates, such as epidote, chlorite and serpentine, often with separation of iron in the form of oxides or hydroxides (magnetite, haematite, limonite), together with carbonates and other compounds of very uncertain character and composition. The weathering of the amphiboles and pyroxenes of the igneous rocks forms a very important source of iron and magnesia in sediments and in soils.

Olivine. This mineral is very unstable under ordinary conditions and readily undergoes decomposition. Olivine is of complex composition, being best described as a mixture of the isomorphous orthosilicates of magnesia, iron and lime. Each of these molecules however undergoes a different kind of decomposition; stated in general terms the magnesia molecule forms serpentine; the iron, which is in the ferrous state, undergoes oxidation and hydration, forming various oxides and hydroxides of iron, while the lime molecule commonly gives rise to carbonate (calcite). Serpentinization is generally attributed to the action of carbon dioxide dissolved in water, the reaction being represented by the following equation:

$$2[2MgO . SiO_2] + CO_2 + 2H_2O = 3MgO . 2SiO_2 . 2H_2O + MgCO_3.$$
$$\text{Olivine} \qquad\qquad\qquad\qquad \text{Serpentine}$$

The change involves a loss of magnesia, since the carbonate is fairly soluble, and also a marked alteration of volume which has a shattering effect on the rock. Cases are also known in which original crystals of olivine have been replaced by dolomite, showing that under certain conditions the magnesium silicate may be wholly changed to carbonate.

The mineral olivine is only found in igneous rocks and by its decomposition it gives rise to compounds of magnesia, iron and lime of some importance as soil-formers.

The weathering of silicates. This subject has been investigated from the chemical side by Van Bemmelen[1]; this author treated soils of different types by successive extractions with acids and alkalies of increasing strength, chiefly hydrochloric acid and caustic soda. He then determined the ratio of silica to alumina in the extracts, in the endeavour to ascertain whether the existence of compounds of constant composition could be demonstrated in the weathered soils. The results were somewhat conflicting, but, after examination of a great number of samples, it was concluded that three distinct types of weathering could be recognized, as follows:

(1) Ordinary weathering, in which the ratio of silica to alumina is approximately 3 : 1. This kind of weathering prevails mostly in moist, temperate climates.

(2) Kaolinitic weathering, yielding products with a ratio of 2 : 1; the most important of these are kaolinite and its amorphous form halloysite, which may be represented by the formula $Al_2O_3 . 2SiO_2 . 2H_2O$. The determining climatic factor is here uncertain.

(3) Lateritic weathering; in soils of this type the chief constituents are hydroxides and hydrates of alumina, such as gibbsite (hydrargillite) and bauxite. Silica when present occurs almost exclusively as unweathered minerals. Lateritic weathering is specially characteristic of tropical regions with distinct wet and dry seasons.

No conclusions could be formed as to the state of aggregation of the bases calcium, magnesium, iron, potassium and sodium in weathered soils. It is quite uncertain whether they exist as crystalline compounds or as colloids, or merely as solutions adsorbed by the aluminous compounds or by the humus. Microscopic investigation of the constituent particles of clays by optical and other methods has revealed the existence of crystalline minerals, especially kaolinite, chlorite, various mica-like substances, silicates of iron and of lime, magnetite,

---

[1] Van Bemmelen, "Beiträge zur Kenntniss der Verwitterungsprodukte der Silikate in Ton-, Vulkanischen und Lateritböden," *Zeits. für anorg. Chemie*, vol. XLII. 1904, p. 265: "Die Verwitterung der Tonböden," *ibid.* vol. LXII. 1909, p. 221.

carbonates of lime and magnesia and others, besides minute unweathered crystal-grains of most of the common rock-forming minerals, especially quartz, felspar and mica. The identification of the mineral constituents of weathered soils is a subject on which much work is needed, but the practical difficulties in the way are great, owing to the minute size of the particles, and the presence of much amorphous material, possibly in the colloidal state, such as the kaolin-jelly of Ramann[1].

The weathering of non-silicate minerals. Many of the minerals which are stable under normal conditions of temperature and pressure are not silicates, but belong to other classes of chemical compounds, some of the most important being oxides, hydroxides, carbonates, sulphates and chlorides. These minerals naturally undergo changes differing much in their nature from those affecting the silicates. Since these minerals are stable under the given conditions the tendency is towards their formation rather than their destruction, and it is necessary to treat them from a somewhat different standpoint.

Magnetite is a stable mineral occurring as an original constituent in rocks of both igneous and sedimentary origin. Under certain circumstances it may be oxidized to haematite, or it may become hydrated and form limonite or other iron hydroxides, but commonly it is passed on as a detrital constituent from one rock to another without change.

Ferrous disulphide in the form of pyrites is a very stable mineral and tends to be produced as a result of low grades of metamorphism, but when existing in the form of marcasite it is very unstable, readily undergoing oxidation to ferrous sulphate and ultimately to sulphuric acid. When abundant therefore it can play an important part in soil changes.

The weathering of the carbonates is a simple process, being for the most part solution in water, generally aided by the presence of carbon dioxide, as explained on p. 47. This process really comes under the heading of erosion rather than of weathering in the strict sense. Rock-salt and gypsum are still more soluble than the carbonates and the same remark applies with even more force.

[1] Ramann, *Bodenkunde*, 3rd edition, Berlin, 1911, p. 245.

Minerals unaffected by weathering. Besides the minerals enumerated in the list previously given, many of the common rocks contain a greater or less proportion of a great variety of other minerals, many of which are not affected to any appreciable extent by weathering agencies or by chemical changes of any kind. Many of these stable species originate in igneous rocks, for example garnet, zircon, rutile and many others. Another group is formed by metamorphism and pneumatolysis of both igneous and sedimentary rocks, e.g. tourmaline, staurolite, kyanite, spinel and again garnet, while a third group includes minerals formed in mineral veins, e.g. cassiterite, and a host of others. Since these minerals are almost indestructible they pass with little change from one rock to another during a cycle of denudation, deposition and cementation; hence they are common constituents of sediments, but usually only in small quantities. From the agricultural point of view they are not important, though many substances of great economic value, such as gold, diamond and many other gems and tinstone, come under this heading. Owing to their high density such minerals often undergo a kind of natural concentration in gravels and sands, and give rise to important industries. Interesting conclusions as to the source and origin of sands and of soils can sometimes be drawn from a study of these mineral constituents, which are easily separated by special methods[1].

[1] Hatch and Rastall, *Textbook of Petrology*, vol. II, The Sedimentary Rocks. London, 1913, appendix.

CHAPTER III

TRANSPORT AND CORRASION

The process of denudation, the destruction of the earth's surface, consists of three stages; the first of these, weathering, is fully described in the last chapter. The second stage is transport and the third, corrasion. The word transport explains itself; it is the removal by gravity, wind, water or ice of the material loosened by weathering. While in movement this material is enabled to do destructive work by virtue of its mechanical energy and this work is summed up in the term corrasion. The geological effects of transport and corrasion are so closely interwoven that it is most convenient to treat them together. Any other course would involve much needless repetition.

As in the case of weathering, transport also is controlled largely by climate and altitude. The chief agents involved in each case may be summarized as follows:

Tropical zone ...	...	Running water.
Arid zone ...	...	Wind.
Temperate zone	...	Running water.
Arctic zone ...	...	Ice.

Gravity is of course equally operative in all areas, and all transport, except by wind, is ultimately due to gravity.

As before pointed out the relative importance of weathering and transport in any given area is largely a question of climate and the consequent presence or absence of vegetation. Transport is at a minimum in the tropics, where the thick vegetation is protective and weathered material accumulates to a great thickness. On the other hand it is at a maximum in regions subjected to considerable variations of climate, where all the

agents have full play in turn. It is obvious also that the configuration of the land must have a great influence in determining
the amount of work done by water and by ice, since the energy
of both of these agents depends on their velocity, and this in
its turn is controlled by the slope. Gravity and water can only
carry material to a lower level; on the other hand wind, and, to
a less extent, ice, can transport material to a higher level. The
general tendency however is nearly always downwards, and the
final stage of transport is deposition in the sea.

The processes of corrasion are somewhat more difficult to
follow, since they are not usually conspicuous when in actual
operation. In general however it is fairly obvious that a river
has carved out its own bed, and the nature of the processes here
involved may now be considered. The simplest of all is solution
and in certain cases this is undoubtedly of importance, especially
when a river runs over limestone rocks. Most corrasion is
however purely mechanical, being due to the kinetic energy of
the water, ice or wind and of the rock-fragments transported
by these agents. Pure water, running over a smooth rock,
without material in suspension, would have little or no erosive
effect, owing to the absence of friction, but when it is carrying
and pushing along grains of sand, pebbles and boulders, these
wear away the rocks over which they travel, acting like a file
or sandpaper. In the case of ice this effect is still more conspicuous, as shown by the scratched and grooved rock-surfaces
seen in glaciated regions. The erosive effect of wind-blown
sand is also well known. Running water and ice also exert
what may be described as a plucking action on the rocks over
which they travel, especially if these are well jointed. Fragments of various sizes, bounded by joints, are thus torn away
from their beds and carried along to perform work in their
turn. A conspicuous feature of water-borne pebbles is their
rounding by mutual attrition, and consequent reduction in
size. Streams also enlarge and widen their beds by undermining the banks, especially when flowing through soft material.

Besides the actual cutting away of the beds of well-defined
rivers and streams there is constantly in operation a general
and slow degradation of the surface of the land due to gravity

and rain.   The effect of rain-wash is shown by the muddy state
of rivers during floods, when vast quantities of finely divided
soil-material are carried away from the land and deposited in
the sea.   The lowering of the general surface of the land is in
the main due to the washing away of soil by rain-water.

Besides all these processes belonging to the land, the geologist
must also take into account transport and erosion by the sea.
This is a subject of much practical interest, especially to land-
owners and farmers near the coast.   Even within historic
times coast erosion has brought about great geographical
changes in the British Isles and elsewhere (see p. 76).

Transport and soil-formation.   These two processes are
mutually interdependent and it is impossible to overrate the
importance of a correct understanding of their relationships.
Sometimes they are antagonistic, in other instances they work
together for a common end.   In general terms it may be said
that the tendency of transport is to take the soil away from one
place and to deposit it somewhere else.   It is evident that there
must be limits to this, but these limits are not actually reached
until the material has been carried far out to sea, beyond the
influence of waves, tides or currents.   Before this condition is
reached any material which has been deposited is always liable
to disturbance owing to varying conditions.

At great elevations little true soil is formed, and in hilly
regions there is always a tendency for the soil to creep downhill
and to accumulate on the floors of valleys and very often also
in lakes, if such exist.   Alluvial flats formed by partial or complete filling up of lakes are a common feature of hilly districts
and they are composed largely of transported soil.   In the
middle and lower parts of the course of a river, where the grade
is less steep, much deposition of soil-material often takes place,
forming wide spreads of alluvium, while deltas and deposits in
estuaries also consist for the most part of material carried
down from the land surface of the upper part of the basin.

From these considerations it is evident that the formation of
soil in any given area depends on the relation between weathering
and transport.   Where weathering is excessive, soil will tend
to accumulate to a great thickness, while the other extreme is

exemplified by some regions, usually high-lying, where transport is so much in excess that the surface consists of bare rock. The maintenance of a constant thickness of soil must be dependent on a delicate balance between the geological processes, and probably under natural conditions the thickness of the soil layer is always either increasing or diminishing, though in most instances very slowly. It is quite evident however that unusually heavy or long-continued rainfall must have a serious effect in removing the finely divided portion of the soil, in all regions except perhaps in perfectly level plains. The sudden and violent downpours commonly known as "cloud bursts," where a great volume of rain is concentrated into a small area for a short space of time, sometimes produce absolutely disastrous effects, removing all the soil and loose subsoil down to the solid rock, and rendering the land useless for agricultural purposes. Cloud bursts are fortunately rare in temperate climates, but in some arid regions they are not uncommon.

Denudation. The combined effects of weathering, transport and corrasion may be summed up in the general term *denudation*; this term in its widest sense implies the general degradation and destruction of the land areas of the world and the transfer of their material to the ocean basins, where deposition comes into play and gives rise to new solid masses, namely, sediments and eventually rocks. The effects of denudation as seen on the land are conveniently expressed by the term *earth-sculpture*. This includes the development of streams and rivers, the carving out of hills and valleys and in short the formation of the surface relief of the land, however this may originate, upheaval and fracture being of course excluded.

In temperate climates, such as our own, rivers are by far the most important agents of denudation and land-sculpture, and it will be well to begin with a consideration of their origin and development. When a new land-area is formed by uplift of the sea-floor it will consist of smooth sheets of sediment inclined in various directions, according to the form of the uplifted area. The rain falling on the land will tend to collect in any accidental depressions of the surface and to run down the steepest slope. This will be in general down the dip of the rocks, and such primary streams may be called *dip-streams*, or *consequent streams*, since they are the direct consequence of the uplift. Another type of consequent stream is formed when the rocks are crumpled up into folds with inclined axes, the streams then running along the synclines. The lower part of the Thames, below Reading, is an example of this. Such longitudinal valleys are very common in mountain regions.

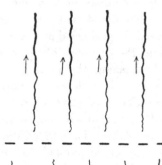

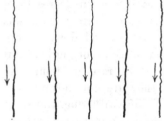

Fig. 18.   Simple dip-streams resulting from uplift.

As time goes on the inequalities of the surface increase owing to the constant removal of weathered material, and subsidiary or tributary valleys develop, often running into the main valleys more or less at right angles; these are called *subsequent* or *strike-streams* (since they are at right angles to the dip streams they must be, by definition, parallel to the strike of the rocks). These again eventually come to possess tributaries of the third order, and so on indefinitely. The final result is a network of streams, such as may be seen on a map of any well-watered country, running generally from the high ground or watershed to the sea, but showing much variation in detail and many minor deviations, owing to differences in the hardness of the rocks over which they run and to other causes.

The hardness of rocks is a feature of fundamental importance in the study of river development, and may be stated as follows: hard rocks tend to stand up as hills and soft rocks to form valleys. This somewhat obvious fact is dignified in America with the title of the *Law of Structures*. Furthermore when a valley is carved out of a series of rocks of varying degrees of hardness, the part of the valley in the hard rocks is narrow and steep-sided, often forming a gorge in extreme cases, while in the softer rocks, the valley is wide and open with gently sloping sides.

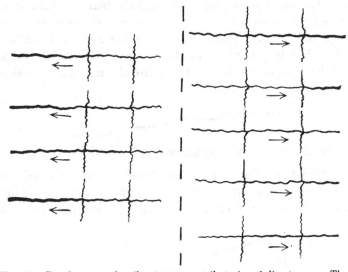

Fig. 19.   Development of strike-streams as tributaries of dip-streams.   The broken line is the primary watershed.

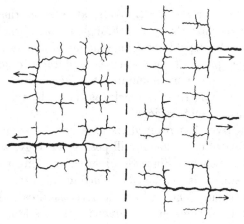

Fig. 20.   A highly developed river system, showing a network of streams: the broken line indicates the primary watershed.

During the earlier stages of its existence a river gradually and continuously deepens its bed, but this can only go on up

to a certain point.   The character of the bed, when considered
in vertical section, depends on the age of the river.   At first
the soft rocks are cut away more quickly than the hard ones,
hence a young river often shows an alternation of steep rocky
gorges and wider stretches with a gentler and more uniform
slope.   Specially hard strata often form rapids and waterfalls,
and the existence of these in the course of a river may be taken
as an indication of youth.

Fig. 21.   Profile of a young river showing inequalities in the grade.

Ultimately all inequalities are levelled down to a uniform
slope and a mature river which can cut down no further in any
part of its course is said to have reached the *base-line of erosion*.

Fig. 22.   Profile of an old river approaching the base-line of erosion.

Probably however no actual rivers, at least in this country,
have quite reached this stage throughout their whole length;
all appear to be capable of some amount of downward erosion
in the upper parts of their courses, and those originating in
hilly and still more in mountainous regions are still actively
engaged in eroding their channels, as shown by the prevalence
of waterfalls in such regions.

When a river has reached its base line throughout a considerable part of its course it still possesses energy, which must
be utilized somehow.   Since the river can no longer cut downwards the energy is employed in cutting sideways, that is, in
lateral corrasion.   The bed of the river is thus widened and
also tends to become more sinuous.   It is seldom that the
course of a river is quite straight and any accidental projection
of the bank on one side will throw the current over towards the
opposite bank, producing an indentation on this side and leaving
slack water on the other side.   In this slack water deposition

goes on, forming first a mud bank or gravel bed, and finally building up alluvial land. This process is cumulative, the bends becoming more and more accentuated as time goes on, while the whole system of curves also tends to travel down stream. In course of time a river may thus work over a very considerable area on either side of its mean channel, producing the large alluvial flats so characteristic of the lower parts of many valleys. These exaggerated bends are called meanders and are often extraordinarily acute; at times the neck of land between two reaches of the river becomes so narrow as to be broken through by a flood and the channel is temporarily shortened and straightened, leaving isolated curved patches of water, such as

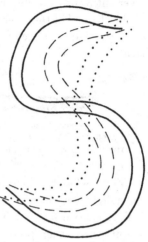

Fig. 23. Diagram to show development of meanders in the course of a river: the curves gradually becoming more and more accentuated.

are called "ox-bows" along the course of the Mississippi. The form of a valley in which meandering has occurred for a long time is quite characteristic, consisting of a flat alluvial plain in

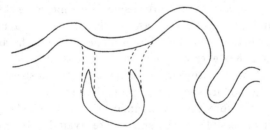

Fig. 24. Crescent-shaped "ox-bow" formed by shortening and straightening of a river bed. The dotted lines show the course before the diversion.

which a winding river flows, while the sides of the valley rise up from the plain as bluffs with a comparatively steep slope. In this way have originated those broad strips of fertile alluvial

land which border the courses of many rivers, such as the
Thames, the Trent and the Yorkshire Ouse. They consist of
river silt down to a depth usually equal to the maximum depth
of the channel of the river itself. This land is usually somewhat
swampy, but when efficiently drained it is often remarkably
rich.

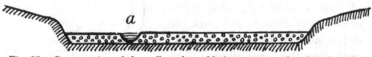

Fig. 25. Cross section of the valley of an old river; a meander plain bounded
by bluffs. The dotted portion represents the alluvium formed by the
river, which is seen at *a*, but has at different times occupied every possible
position within the limits of the meander belt.

From what has been said above it will readily appear that
the relief of a surface produced by river denudation depends
mainly on two factors; the length of time during which earth-
sculpture has been proceeding and the original structure of the
area in relation to the varying hardness and character of the
rocks composing it. No doubt the general lowering of the land
is an extraordinarily slow process; it has been estimated on
what appear to be fairly reliable data at an average of 1 foot
in 6000 years. Such a rate will obviously produce no visible
effect during an ordinary lifetime, and it is difficult or impossible
to establish the existence of any real change, even from the
oldest historical records. Geological time however is unlimited,
and in theory at any rate, given sufficient time, even the highest
mountains should be worn away. It is however a matter of
dispute whether land-denudation alone, without the assistance
of the sea, has ever been able to reduce a hilly country to a
plain. It would appear probable that in most cases some
disturbing factor, such as a general uplift or subsidence, is
introduced before this consummation can be reached.

It is impossible in the space here available to give a full
account of the important subject of river development and of
denudation by water-action. Of late years the theory of
denudation has been extensively developed in America, where
the geological structure is simpler and the conditions somewhat
more favourable than in Britain, with its marked local variations

of climate and great complexity of geological structure. In the British Isles also ice-erosion has played a considerable part in modifying the relief of the land. At the present time the geological effects of ice and water have not been satisfactorily discriminated in this country, and in many cases there is still much dispute as to the precise nature of the principal agent concerned in the production of many of the prominent features of our valley systems[1].

Water denudation and topography. In regions of temperate climate, where the effects of ice are excluded, the form of the land surface is mainly due to the action of running water. Owing however to variations of original rock-structure the forms produced show endless diversity, depending on relative hardness of strata, inclination and folding of the rocks, faults, landslips and many other factors. It may be said in general terms that the broad outlines of the distribution of the land and the arrangement of the principal mountains and watersheds depend primarily on earth movement, while the details of coastal forms are mainly due to the sea, but the minor features of the land, the hills, valleys, cañons, gorges and waterfalls are determined by streams and rain.

In a region where denudation has been going on for a long time the valleys are wide and open and the outlines of the hills rounded, smooth and flowing. Serrated ridges and narrow rocky valleys indicate a youthful drainage system, where time has not yet sufficed to wear down the asperities of the surface. Waterfalls in the course of a river are almost always indications of youth; in old rivers and in those at and near base-level the course is worn down to one uniform curve. Coming to the details of rock-erosion in actual river-beds, special importance attaches to *pot-holes*. These are circular hollows produced in the rock by the gyrations of stones in eddies; they are often very deep and undoubtedly play an important part in the scooping out of many river-beds, where the current is rapid and pebbles abundant.

---

[1] For a concise account of the geological effects of running water see Bonney, *The Work of Rain and Rivers*, Cambridge Manuals of Science and Literature Cambridge, 1912.

The term cañon is one which has been much abused. In America it is now used to describe almost any kind of valley, but in this country it generally connotes the idea of a deep and narrow chasm in the earth; such would be perhaps better designated a *gorge*. Such deep, narrow valleys are formed in two principal ways: firstly, if the rocks through which the river runs are very hard and resistant, the denudation of the surrounding slopes is small and the sides of the valley remain steep; secondly, if a river running from a normally wet region cuts through a dry area, denudation of the surrounding country is again at a minimum and the same result follows. Some of the greatest gorges of the world, such as the Colorado Cañon, are due to a combination of special causes. In the first place the Colorado river flows from a mountain region into a dry, almost desert tract, but furthermore, after the river had established its course and reached base-level, an uplift of the whole country took place, so that the river was enabled again to cut deeply into the strata. The gorge is now some 300 miles long and has a maximum depth of about 6000 feet.

In some cases rivers flow straight towards mountain ranges and cut through them in narrow deep gorges. In such an instance the only possible explanation seems to be that the river is actually older than the mountains, and was strong enough to keep open its original course, as the mountain chain rose under the influence of crumpling of the crust. Such is the common explanation of the Iron Gate of the Danube and the valleys of the Indus and the Bramaputra, both of which rise on the north of the Himalaya and cut right through the range.

It would be easy to go on multiplying indefinitely instances of peculiar topography directly due to river denudation under special conditions, but space will not allow of any further amplification of the subject. Enough has been said to bring out the main points at issue. For a detailed treatment reference must be made to some general text-book of geology.

Denudation in limestone regions. Owing to the ready solubility of calcium carbonate in water, the results of water-erosion of limestones present certain peculiar characteristics. Besides being soluble, limestones are usually very well jointed.

The joints are produced in the first place by shrinkage, but they are rapidly widened, often to considerable depths, by solution, and consequently any rain falling on, or streams running over, the surface soon disappear down these open joints; in limestone countries most of the water circulation is underground and the surface remains dry and waterless, often consisting merely of bare rock. Thus is produced that topographical type, perhaps most highly developed in the region known as the Karst, to the east of the Adriatic, hence known as the Karst type of scenery. It is also very well seen in many parts of the British Isles where the Carboniferous Limestone forms the surface rocks, as in west Yorkshire, north Lancashire, Westmorland, Derbyshire, the Mendip Hills and elsewhere. Highly characteristic are the *grikes* and *clints* of west Yorkshire, Lancashire and Westmorland, elevated plateaux of limestone, consisting of bare rock, with many wide open joints, in which alone plants find sufficient moisture for their existence. A very fine example of this is seen on Ingleborough, and this region is also remarkable for its caves and swallow-holes, both features highly characteristic of limestone regions. The formation of caves depends on the widening of joints and fissures by solution, aided by the collapse of undermined portions of rock. Very commonly an underground river runs through a cave, often falling into it down a swallow-hole, as at Gaping Ghyll on Ingleborough. These swallow-holes are vertical shafts, sometimes two or three hundred feet deep. They originate as vertical joints, and are enlarged and deepened by stream-erosion. When a limestone plateau is covered by a thin layer of drift or peat, swallow-holes may be formed in the limestone beneath and the superficial layer sinks into them, forming conical hollows; these, when numerous, give the surface a curious pitted appearance. Similar structures are sometimes formed in the Chalk.

The topography of Chalk areas is of a somewhat special type and, owing to its wide occurrence in England, worthy of some description. Chalk is a limestone and therefore soluble in water. Its drainage is consequently for the most part subterranean, as in other limestone areas. But it is likewise soft and therefore, except in sea-cliffs, it does not form rocky scarps like the Carboniferous and other hard limestones. Chalk hills, like the Downs and the Chilterns, are consequently smooth and undulating in outline, though often fairly steep. Surface streams are generally absent; nevertheless a well-developed valley system is generally to be seen. This is somewhat difficult to account for, but is generally explained on the supposition that the valleys were formed when the ground was frozen to a considerable depth, during the glacial period. Then the streams were compelled to run on the surface, being unable to sink into the frozen rock, and thus carved out the valleys in the usual way. A similar explanation will account for the dry limestone valleys of the north-west of England. The origin and characters of the soils found lying on the Chalk are discussed elsewhere (see Chapter XIV).

Denudation in arid regions. In those parts of the globe where the rainfall is deficient the geological conditions are quite special and of a peculiar character. This type of denudation is not found in the British Isles, and is only seen in its full development within the desert belt. Roughly speaking an arid region may be defined as one where the average annual rainfall is less than 10 inches. Such conditions prevail for example over a large part of both north and south Africa and in Australia. The day temperature is very high and, owing to the clear skies, there is a rapid fall after sunset. Hence changes of temperature are very sudden and produce important effects by alternate expansion and contraction of the rocks, causing conspicuous shattering. This is the principal weathering agent, and there is comparatively little chemical action, owing to the absence of water and vegetation. The fragmental material is therefore very fresh and undecomposed. The chief agent of transport is wind, though gravity plays an important part in mountainous regions. Under such conditions sand is formed in enormous quantities and transported for long distances by wind. While in movement the sand performs a certain amount of work in erosion, but according to the best authorities the actual amount of denudation due to sand is generally insignificant, though its superficial effects are striking, consisting of rounded and polished surfaces. In regions where desert topography is fully developed, agriculture is impossible and the subject need not concern us further. There are however many cultivated regions, sometimes comparatively fertile, where the rainfall comes below the limit specified, e.g. parts of the Union of South Africa and of Australia. Here for part of the year the climate is very dry and for many months no rain may fall. The greater part of the rainfall is concentrated into a short season, and the downpours when they do occur are often violent. Hence the effects of denudation are somewhat complicated. In the dry season denudation of the arid type is dominant, while in the wet season water action is conspicuous and floods often produce very well-marked effects. In such regions geological processes are on the whole probably more rapid than in temperate climates and the topography is highly accentuated, as in the mountain regions of Cape Colony.

Ice as an agent of denudation. Here we touch upon one of the most difficult and controversial of all geological subjects. Of the potency of ice as an agent of transport there is no question ; it is obvious on the most cursory view of a glacier or of an iceberg, but when we come to the consideration of erosion by ice, it is another matter. This is a subject on which the most contradictory views are held by authorities of the highest rank.

In the high mountain regions where glaciers are found, and in the arctic lands, the chief agent of rock-disintegration is frost; fragments fall from the mountain slopes on to the ice, and are carried along by it, either on the surface or within the body of the ice. These form the accumulations known as *moraines*. As the ice moves slowly downwards the stones are carried along with it, to be deposited eventually at the point where the ice melts, forming the terminal moraine. This is usually a crescent-shaped mound, running across the valley in which the glacier terminates, and consisting of more or less angular fragments derived from the surface moraines of the ice ; but the terminal moraine itself includes a good deal of finely divided material, and the streams running from the end of the ice are always turbid with fine mud, the so-called *rock-flour*. The question next arises what the source of this may be.

Part of it is certainly formed by the mutual attrition of fragments, but this cannot account for much and we are driven to conclude that the glacier really does erode its bed as a whole, mainly by means of stones embedded in its lower surface and dragged along, though also doubtless to a considerable extent by plucking, i.e. by tearing away of blocks along joint planes.

No one doubts at any rate that glaciers have an important and conspicuous effect in modifying the surface over which they move. This is shown by the rounding, polishing and grooving of projecting rocks, in any region where glaciers have recently existed; the main controversy is as to whether the erosive effect of ice is purely superficial, merely putting a final polish on to features already formed by water action, or whether ice-erosion has been the principal agent of land-sculpture in regions where ice has once existed. The question cannot be answered yet, since the evidence is too conflicting, and it is not of much interest to the agriculturist, who is concerned more with the effects of ice as an agent of deposition.

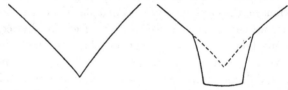

Fig. 26. On the left a V-shaped valley due to water-erosion: on the right a valley with a U-shaped section in the lower part, due to deepening by ice of a river valley. The dotted line shows the form of the original river valley.

It is generally believed at the present time that the form of the cross-section of a valley gives some indication of the nature of the agent by which it was formed. A valley with uniformly sloping sides and very narrow at the bottom, such as is best described by the term V-shaped valley, is supposed to be due entirely to water; such valleys are often winding as well as steep and narrow and projecting spurs are conspicuous. On the other hand a U-shaped valley is supposed to be due principally to ice; here the sides are very steep, often indeed precipitous, but there is a fairly flat floor, often of quite considerable width.

U-shaped valleys are generally straight and the tributary
streams fall into them abruptly, often as waterfalls; projecting
spurs do not exist, having been planed away by the ice.

Evidences of the former existence of glaciers are conspicuous
in most of the mountainous regions of our own country, especially
in Wales, the Lake District and Scotland. They consist of the
characteristic polished and scratched surfaces before mentioned,
together with *roches moutonnées* and perched blocks. A *roche*

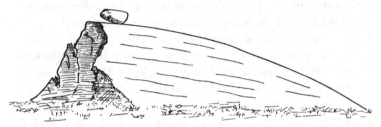

Fig. 27. A *roche moutonnée*, with a perched block resting on it. The ice
moved from right to left.

moutonnée is a projecting knob of rock that has been smoothed
and polished on the side from which the ice came, while the
sheltered side is left in a rough and jagged condition; perched
blocks are masses of rock left lying by the ice on the surface
of the ground, often after having been transported for long
distances. The general term boulder is often employed to
connote masses of rock of considerable size that have been
transported from a distance and either left lying on the surface,
or embedded in gravel, sand or clay. Besides the foregoing
there are often to be seen moraines of all kinds, whose origin is
indicated by their characteristic form, and also many accumulations of gravel, sand and other material, either deposited
directly by the glaciers or by the streams running from them.
Even the lowland tracts of Great Britain north of the Thames
are for the most part covered by a sheet of glacial deposits,
largely consisting of boulder clay. These superficial deposits
are of enormous agricultural importance and their characters
are fully described elsewhere (see Chapters v and xvi). They
are only mentioned here as physical evidence for the former
existence in this country of glacial conditions.

Marine denudation and coast-lines. This is a subject of
much practical significance in certain maritime regions, although
to the inland agriculturist it is not a matter of concern. It has
long been known that great changes have occurred along the
coasts of the British Isles and of late years the subject has been
considered so important that a Royal Commission was appointed
to carry out a thorough investigation. The final report of this
commission shows that while there has been much loss of land
by marine erosion in certain districts, it is more than counterbalanced by gain in other localities, so that the total area of
the British Isles has slightly increased in recent years. However it is poor consolation for a landowner in Yorkshire or
Norfolk who has lost a whole farm, to know that some one in
South Wales or Lancashire has gained a similar or larger area.
In general it may be said that destruction is most rapid on the
projecting parts of our eastern coasts, in Holderness, Norfolk
and Suffolk, while the gain is taking place chiefly in the long
estuaries and narrow inlets of the west coast of England and
Wales[1]. It is estimated that during the last 35 years the loss
has amounted to about 6600 acres, while the area gained is
estimated at 48,000 acres. Much of this new land however is
as yet of little or no agricultural value, being still largely mud
flats and salt marshes, which in course of time will doubtless
become good, fertile land.

Marine denudation is due to three principal causes; waves,
tides and currents; these usually act in conjunction and their
effects are often separated with difficulty. The efficiency of
these causes naturally depends primarily on the character of
the rocks on which they act. It may seem a platitude to say
that soft rocks are eroded more quickly than hard ones, but
nevertheless this simple statement is the main foundation of
the whole subject. Where hard and soft rocks alternate the
coast-line is likely to be indented, especially if exposed to the
full fury of the waves, as in the south-west of Ireland. On the
other hand rocks of uniform hardness tend to give rise to smooth,
flowing shore lines, as on the east of England, especially in

[1] *Final Report of the Royal Commission on Coast Erosion*, London, 1911,
p. 158.

Northumberland, Lincolnshire and Norfolk. The subject is however often complicated by submergence or emergence of land. The former tends to form an indented coast, since the sea flows up river-valleys, as in the west of Scotland, while emergence lays bare a smooth area of sea-floor, forming a flat coastal plain of simple outline, as in the eastern United States. Deposition also tends to fill up bays and estuaries and to reduce inequalities of outline. Thus the Wash, once a deep gulf running up nearly as far as Cambridge, was subsequently for the most part filled up by deposition of sediment. The origin of the Fenland is discussed elsewhere (see p. 128).

The most serious coast-erosion now taking place in the British Isles is seen on the Yorkshire coast between Flamborough Head and Spurn Point, and on the coasts of Norfolk and Suffolk, from Cromer southwards. In both these areas the cliffs consist of soft glacial deposits, boulder clay, sands and gravels, which are easily washed away by the strong southward set of the tidal currents. It is estimated in the report of the Royal Commission before quoted that the loss of land in Holderness in a period of 45 years amounts to about 770 acres, and in East Anglia in a little over 30 years nearly 1000 acres have been swept away[1]. In both areas the sites of well-known mediaeval towns and villages are now far out to sea.

Coast-erosion is mainly due to two causes; actual friction on the rocks of the shore, caused by the transport of stones and shingle by the waves, and landslips due to undermining at the base of the cliff. In hard and well-jointed rocks, especially those with open joints, great importance is to be assigned to the sudden compression of air within the cavities of the rock, when waves dash up against them. This cause is specially operative when cliffs descend into deep water, with no beach at the base. In such cases lines of caves are often formed somewhere about high water mark, extending far inland. This also leads to undermining and falls of cliff. Caves are specially common in limestone, and especially in Chalk cliffs,

---

[1] *Op. cit.* p. 43. The figures for East Anglia are generalized from the data given for separate counties, with slightly varying dates in each case.

as at Flamborough Head and on the Antrim coast and in the Carboniferous Limestone in South Wales.

While wave-action is mostly confined to the lower part of the cliff, ordinary denudation by rain and frost is acting on the upper part, and wearing it away. Hence the steepness of the cliff depends on the ratio between these two kinds of denudation.

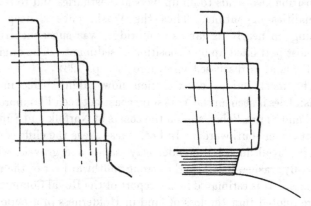

Fig. 28. On the left, a cliff in well-jointed rock of uniform texture; on the right an overhanging cliff formed of hard rock above with a soft layer below.

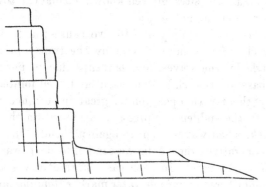

Fig. 29. A cliff of well-jointed rock with a platform at the base due to wave-action, forming a beach.

If the cliff is composed of one kind of rock from top to bottom, a hard rock will form a steep cliff, owing to the small effect of subaerial denudation, while a soft rock will be weathered away more quickly above, forming a gently sloping cliff; in fact in many places where the rocks are all soft there is no cliff at all, but the surface of the land slopes down evenly and continuously to the beach. Some of the steepest of all cliffs are formed where a hard rock overlies a soft one; these cliffs may even overhang, since the base wears away more quickly than the upper part.

For our present purpose marine deposition is of far greater importance than denudation, so that all consideration of the nature of sea-beaches may be postponed for the present. The nature of coast deposits of all kinds is discussed fully in Chapter IV.

CHAPTER IV

SEDIMENTS

The sedimentary deposits in general. It has already been explained that the modern sediments and the ancient rocks of similar character have been formed either directly or indirectly from the materials of pre-existing rocks, and that the primary source of all this material is to be sought in the original primitive crust of the earth, this being of necessity of igneous origin, or at any rate formed at a high temperature. Consequently the sediments are partly composed of the same minerals as the igneous rocks, in an unaltered state, and partly of the products of the weathering and decomposition of these minerals.

Sediments may be formed in a large number of different ways and the resulting products show much variation in character and composition. As a matter of practical convenience they are generally treated under three headings according to their manner of formation, namely, *mechanical, chemical* and *organic.*

The *mechanical sediments* include those deposits formed as a result of denudation and deposition by running water, ice, wind and gravity; the *chemical sediments* are formed by evaporation of solutions, precipitation and other similar processes of purely physical and chemical nature, while the *organic sediments* are directly due to the vital activity of animals and plants.

An alternative method of classification, and one perhaps more strictly scientific, is framed according to the conditions under which the deposit is formed, namely, marine, estuarine, fresh-water and terrestrial. This classification, however, if applied in detail, involves much repetition, since very similar deposits are formed under different conditions.

The mechanical sediments. The subdivisions of the mechanical sediments, according to the classification here adopted, are founded on the size of the component fragments. Any division of this kind must of course be purely arbitrary and all possible gradations may exist. As a basis for the groups are taken the popular and generally understood terms, boulders, pebbles (or gravel), sand and mud. The limits of size are approximately as follows:

Boulders are more than 4 inches (10 centimetres) in diameter, i.e. larger than the human fist.

Pebbles vary from 4 inches to the upper limit for sand, about one-tenth of an inch.

Sand includes grains smaller than this and yet distinctly visible to the unaided eye.

Mud (or *dust* when dry) is composed of microscopic and ultramicroscopic particles.

The above is the nomenclature applied to the modern deposits of this group, and the older consolidated sedimentary rocks have been formed from similar deposits by various processes of hardening and cementation, brought about by pressure and chemical agents. The finest types of sediment, dust and mud, may apparently be consolidated by pressure alone, induced by the weight of the overlying strata, but the coarser sediments can only be welded into solid rocks by actual deposition of material in their interstices. This material is for the most part deposited from solution in percolating water, being largely derived from the zone of weathering near the surface and carried down by the underground circulation, but also in some cases brought up from great depths by heated waters, perhaps of volcanic origin.

The chief substances which thus act as cements are silica, calcium carbonate and various oxides and hydroxides of iron, other less common cements being gypsum and barium sulphate. Sometimes the cements are deposited in the amorphous form, as in the case of many of the iron compounds and opaline silica, but more commonly they are crystalline, e.g. quartz, calcite, and gypsum. Sometimes finely divided interstitial muddy matter acts as a cement for the larger elements of the rock, but most commonly there has been some infiltration as well.

The coarser types of mechanical sediment, when cemented, form conglomerate if the boulders or pebbles are rounded, and breccia if the fragments are angular. Hence conglomerates are formed from rounded and water-worn boulders, pebble deposits and gravel, such as those of sea-beaches and of rivers, whereas breccias are chiefly composed of scree-material and similar accumulations due to frost action or to dry weathering in arid regions. The pebbles and fragments composing them may consist of any kind of rock and no distinctive names are in use for varieties differing in this respect. The cement also may be of any kind.

The forms assumed by boulders and pebbles show some variety according to the conditions of formation. Those due to water action are for the most part well rounded, sometimes almost spherical, but more commonly ovoid or ellipsoidal. This shape is due to a combination of rolling and sliding movements. The fragments composing glacial deposits are generally angular and often scratched, while rock fragments that have been acted on by wind-blown sand are polished and faceted, and sometimes worn into strange shapes. In residual deposits due to weathering of rocks in place, the fragments may have any shape whatever, but they are commonly more or less rounded. The most sharply angular fragments are found in screes, which are due to weathering by frost action or changes of temperature, followed by accumulation of the material at the foot of steep slopes in mountain regions. In considering the forms of pebbles in relation to conditions of formation it must always be remembered that the pebbles may have been derived already rounded from some older deposit, so that they are more rounded than they ought to be. Many pebbles of specially resistant rocks have certainly been handed on from one deposit to another through several geological cycles.

Conglomerates and breccias. As before stated these are formed by cementation of boulder and pebble deposits of ancient date. These are abundant among many of the rock formations of all ages composing the geological series.   A few representative examples may be described briefly.   Since pebble deposits are usually formed either in shallow water, salt or fresh, or actually on land, they are very commonly associated with unconformities and breaks in the stratigraphical succession.   During both submergence and emergence of land, pebble beaches are formed, and at the base of a series of marine strata there is usually a conglomerate resting on the denuded surface of the older rocks, while the sediment becomes finer upwards.   Such is the case, for example, where the marine Cambrian strata rest on the pre-Cambrian volcanic rocks in parts of Wales and western England.   Similar conglomerates are also seen at the base of the Carboniferous in west Yorkshire, and in many other instances too numerous to mention.   Figs. 16 and 17 show two different ways in which the basal conglomerate may be related to the rocks above and below.   Among the older rock-systems are also found conglomerates that appear to be of terrestrial origin, having been formed by rivers.   Well-known examples are the conglomerates in the Old Red Sandstone of the north of England and of Scotland, and the Bunter pebble beds in the Lower Trias of Devonshire, Staffordshire and Cheshire; these are often very imperfectly cemented, easily undergoing disintegration to form loose gravels again.   The "Hertfordshire Pudding Stone" of Tertiary age is a very remarkable conglomerate, consisting of rounded pebbles of flint in a matrix of quartz sand cemented by quartz.   One of the best known examples of a breccia is the "Brockram" of Permian age, found in Cumberland and Westmorland; it consists of angular fragments of Carboniferous Limestone, more or less converted into dolomite and embedded in a red marly substance with crystals of gypsum. It was probably formed in a salt lake at the foot of a steep slope, under desert conditions.

Sands and sandstones.   The rocks formed by cementation of sands show a good deal of variation in character; the most typical member of the group is naturally sandstone; the meaning of other terms employed here will be explained shortly.   The most common cementing materials in the sandstones and other rocks of this group are silica, oxides of iron and calcium carbonate; gypsum and barytes as cementing materials are so rare as to be negligible from the agricultural point of view. The cements of sandstones are almost always deposited from solution in percolating water, and the hardness of the resulting rock naturally depends on the nature of the cement and the extent to which the interstices between the grains are filled up. The softest and least consolidated sandstones often possess a cement of limonite or some other iron oxide, deposited only at the points where the grains are actually in contact, leaving considerable interstitial spaces; frequently in sands formed under arid conditions each grain is covered by a skin of red ferric oxide, and this helps to bind the sands into a more or less coherent rock. In many soft estuarine and marine sandstones the cement is some form of brown hydrated iron oxide, e.g. limonite, and this occasionally becomes quite hard, giving rise to *ferruginous grits*. Sandstones with a calcite cement are usually fairly hard, since the calcite is commonly deposited in the form of large crystals, each enclosing many sand grains, and showing a characteristic lustre-mottling when broken; such rocks are often called *calcareous grits*. The hard sandstones of the older rock-systems generally have a cement of silica in the form of quartz, or sometimes a mixture of quartz and finely divided mica, the latter being formed by decomposition of felspar. A hard rock of compact texture composed entirely of grains of quartz with a cement of the same mineral is generally called a *quartzite*, though some writers restrict this term to a rock which has been recrystallized by heat or pressure (i.e. a metamorphosed sandstone).

The word *grit* is very often used as a descriptive term and a good deal of difference of opinion exists as to its proper application. Generally it appears to mean a sandstone of somewhat coarse and uneven texture (e.g. the Millstone Grit of the Carboniferous system) characterized by a rough, gritty feeling to the touch; this might otherwise be designated a pebbly sandstone. An *arkose* is a sandstone or grit with abundant fragments of felspar, while *greywacke* is a somewhat old-fashioned term applied to the grey or greenish sandstones and grits so common in the oldest sedimentary formations.

In many of these the prevailing green colour is due to the presence of much finely-divided chlorite in the cement, probably derived from alteration of mica and other ferromagnesian minerals.

The colour and general appearance of a cemented sandstone depends very largely on the character of the cementing material; generally speaking quartzose and calcareous sandstones are white, pale grey or yellowish, while ferruginous sandstones are yellow, brown or red.

Sandstones, especially those formed in a shallow sea, often contain abundant grains of glauconite, a mineral of somewhat variable and indefinite composition, being mainly a silicate of potash and ferrous iron. The presence of much glauconite gives a distinctive green tinge to the rock. It is perhaps best known in the "Greensands" of the Cretaceous (see Chapter XIV), but glauconite is found in marine sands of almost all ages. Sands rich in glauconite are often associated with important phosphate deposits, as will be explained later, and the presence of this mineral is a useful indicator for the possible presence of phosphate beds. Even if phosphate is not found in appreciable quantity it is improbable that a soil derived from a rock rich in glauconite will be seriously deficient in phosphoric acid, unless this constituent has been exhausted by crops.

Many sandstones are rich in flakes of mica, and a large proportion of this mineral has considerable influence in determining the suitability of the stone for various economic purposes. Flakes of mica tend to accumulate in layers, along the bedding planes and highly micaceous sandstones show a strong tendency to split into flat slabs, which are commonly known as flags. Many other well-bedded sandstones, without much mica, are often called by the same name, and some of the flags of commerce are coarse gritty slates or even finely crystalline mica-schists, both the latter groups being properly metamorphic rocks. The best flags for paving purposes are found in the Coal Measures of west Yorkshire and in the Old Red Sandstone of Caithness, but they are now largely superseded by various artificial stones, made of granite chips and Portland cement, cast in moulds to the required shape and thickness.

As before stated, in nearly all sands and sandstones quartz is by far the most abundant mineral; next come felspar and white mica. Among the less abundant minerals various compounds of iron, such as magnetite, ilmenite and limonite are perhaps the most important. Most sands also contain a small proportion of other hard and heavy minerals, such as garnet, tourmaline, hornblende, augite, kyanite, staurolite, rutile, zircon and many others. These are for the most part of scientific interest only; a few of them, such as apatite, may by their decomposition yield plant food in soils, but most are very resistant to weathering agents. Many recent sea-sands are very rich in shell fragments, which may ultimately yield lime; they are often dissolved out before cementation begins. Sands and sandstones of ancient date are as a rule not rich in fossils.

Concerning the chemical composition of the sandy deposits, there is little to be said. Since quartz is almost always the chief constituent the proportion of silica is necessarily very high, and some fresh uncemented sands, siliceous sandstones and quartzites, are almost pure silica. The presence of felspars and micas introduces alkalies and alumina, while the other minerals mentioned above contain magnesia, iron, lime and other constituents. But the proportion of all these is low in the purer sandstones, rising in amount in the impure varieties, such as arkose and greywacke. The arkoses in particular sometimes approximate in composition to igneous rocks, and by their decomposition may be valuable sources of potash, phosphoric acid and lime. However the plant food in sandy rocks and in sandy soils is generally in a form not readily available, and pure sands, whatever their actual chemical composition may be, are commonly almost barren. The reason is mainly physical, being closely connected with the comparatively large size of the particles and the consequent low water-holding capacity.

The nature of the cementing material must of necessity have a strong influence on the bulk composition of the rock. If this is ferruginous or calcareous the analysis will show a high proportion of iron and lime respectively, with a corresponding effect on the nature of the soil.

Muds, clays and shales. These, which are known collectively
as the argillaceous deposits, represent various degrees of alteration of the finer grained deposits, laid down in the sea in fairly
deep water, at some distance from land, or, under special
conditions, in shallow water or even on land. In general terms
however it may be said that the argillaceous rocks are generally
marine. They are formed of the finest material derived from
denudation of the land, and owing to the minute size of the
particles, the exact mineral composition is often a matter of
uncertainty. When minerals are recognizable under high
magnification they are commonly the same as in the sands, but
there is generally present, in addition, more or less of a structure-
less jelly-like substance, the so-called "colloidal clay." As to
the nature and even the real existence of this colloidal clay
there has been much dispute (see p. 57).

The marine muds are generally grey or bluish in colour; red
and green muds are also known to exist in the sea, but they are
very local and limited in their distribution. The prevailing
grey tint is commonly due to finely disseminated iron sulphide,
with sometimes carbonaceous matter in addition, derived from
the decay of animals and plants. When partly consolidated by
pressure and drying, and subsequently uplifted to form land,
the blue and grey muds give rise to the grey clays which are so
common in many districts in the midlands and in the south of
England. A further degree of hardening converts them into mudstones, or if laminated, into shale (for a discussion of the origin of
lamination see p. 16). When subjected to cleavage these rocks
form slates; the latter are strictly speaking metamorphic rocks.

The red clays and so-called "marls" of the Trias and other
formations were produced in a different way. They consisted
originally of fine dust transported by wind during the prevalence
of arid conditions and either piled up in hollows on the land
surface or sometimes deposited in salt lakes. This accounts
for their frequent association with beds of rock-salt and gypsum,
as in Cheshire and Worcestershire. The red colour is due to
the absence of any organic matter to reduce the iron from the
ferric to the ferrous state, and may be taken as an indication
of deposition under desert conditions.

Muds are very commonly formed on a large scale in the estuaries of rivers, owing to the coagulating effect of the salt water on the finely-divided muddy material brought down by the river, leading to rapid precipitation. Estuarine muds give rise to clays and shales much like those of marine origin, but distinguishable by the nature of the included fossils, which comprise a mixture of marine and fresh-water or even terrestrial forms.

The deposits in large fresh-water lakes generally include beds of mud and clay; these often contain a considerably higher proportion of lime than the muds of the sea, and many of them are more properly designated *marls*, in the true sense of this term (see p. 93). These when consolidated form calcareous mudstones or very impure limestones.

Terrestrial accumulations belonging to this class are mostly formed by wind, a notable example being the Loess of central Europe and Asia. This is a fine brown loam without stratification and often attaining a thickness of hundreds or even thousands of feet, as in China. It is believed to consist of dust of glacial origin, redistributed by wind during warm interglacial episodes. When modified by addition of organic matter it forms the famous "black earth" of Russia (see p. 148).

The chemical composition of the argillaceous rocks shows a wide range of variation, but nearly all varieties are distinguished by a fairly large proportion of alumina, this being the most characteristic constituent, while silica generally averages about 50 per cent. Owing to the presence of much finely-divided felspar and mica, the clays are commonly rather rich in alkalies in a fairly available state, and clay soils are rarely deficient in potash, though lime is often in small amount. A very important property of clays is their power of taking up salts from solutions that come in contact with them. This property, which is known as *adsorption*, is no doubt of great significance in the study of the fertility of clays and loams. It seems to be a purely physical process and not true chemical combination. The composition of clays is much affected by the manner in which they have been formed and the amount of chemical change undergone by the mineral constituents during the preliminary weathering. The most highly altered are the so-called *residual clays*. These are the insoluble residues of highly weathered rocks and are often of very peculiar composition, some being even derived from limestones. The calcium carbonate has been leached out by water and the clay is nothing but the muddy material that was originally enclosed in the limestone at the time of its formation.

The other extreme is shown by the clays of glacial origin. These are formed almost entirely by mechanical attrition, the material having undergone little or no chemical change. Consequently there has been no leaching and all the original constituents of the rock may be present in a comminuted form. Such clays will obviously vary in composition according to the character of the parent rock, which may be of any kind.

Ordinary marine and estuarine clays are intermediate in character; some of the material consists of fresh mineral chips of minute size, loosened by erosion, while another part is a decomposed residue of weathered minerals. In many ancient clays and shales and in all slates there has been also a development of new minerals, largely of a micaceous nature, formed by chemical reactions between the original constituents.

Some special varieties of clay are worthy of brief mention:

China-clay is a product of the weathering of the felspars in granite; its formation has already been described (see p. 52).

Pipe-clay is very similar in appearance to china-clay, being pure white; it is probably also very similar in origin.

Fire-clay is a variety with an unusually low proportion of alkalies. Bricks made of this clay consequently resist heat well and are used for lining furnaces. Beds of fire-clay often occur immediately underneath coal seams. These are supposed to be in many cases the soil in which the coal-forming plants grew, the absence of alkalies being due to the growth of the plants.

Fullers' earth is a peculiar clay which does not become plastic but crumbles away when wetted. It is often rather rich in magnesia, lime and soda. It possesses the property of removing grease from cloth, and is also used for the purpose of decolourizing dark oils. It is found in the Jurassic and Cretaceous systems.

The limestones. This group includes an enormous number of rock-types, very variable in character and origin, but all agreeing in containing a large proportion of calcium carbonate. Limestones may be of mechanical, chemical or organic origin, and may be formed under almost any set of physical conditions. The greater part are marine, but they may be formed in fresh water or even on dry land, though this is not common. Owing to the solubility of calcium carbonate in water the denudation of limestone regions is of a special character and limestones sometimes give rise to very peculiar soils, or even on occasion to no soil at all, the surface being bare rock.

The mechanically-formed limestones are sediments in the true sense of the word. They consist of the debris of older limestones, loosened by weathering and transported to an area of deposition. Some of them, such as the limestones of the Lias in Dorset, are consolidated calcareous muds, derived probably from the Carboniferous and older limestones, carried down by rivers into the sea and there deposited, to be subsequently consolidated by pressure and drying. The Cornstones of the Old Red Sandstone system of Hereford and South Wales are best described as calcareous breccias on a small scale, formed in a very similar way. It seems probable from the results of recent research that limestones formed in this way are more common than was formerly believed. It is often difficult to establish any clear distinction between limestones of this class and those formed by broken fragments of calcareous organisms, as will shortly appear.

The chemically formed limestones originate chiefly by precipitation from saturated solutions. When calcium carbonate is acted on by water containing carbon dioxide a bicarbonate is formed, thus:

$$CaCO_3 + H_2O + CO_2 = CaH_2(CO_3)_2$$

and this salt is more soluble in water than the normal carbonate. Since the amount of carbon dioxide that water can hold in solution is increased by pressure, underground waters are often rich in the bicarbonate. When the water comes to the surface as a spring the salt decomposes, the gas escaping into the air;

the normal carbonate is again formed and is precipitated in the solid form. Thus originate the great deposits of travertine and calc-sinter which are so noteworthy a feature of some limestone regions. The stalactites and stalagmites of caves originate in a similar way.

Another interesting example of chemically formed calcareous deposits is afforded by the "surface limestones" of South Africa and other hot, dry countries. Owing to surface evaporation water rises by capillarity from below, bringing with it calcium carbonate in solution. This is deposited in the interstices of the soil and rocks immediately below the surface, resulting in the formation of a layer of calcareous matter, which is often very hard, and is an effectual bar to cultivation[1].

Many limestones of various ages, and especially those of the Jurassic system (see Chap. XIII) are characterized by possessing what is known as *oolitic* structure. Such rocks show a strong resemblance to the roe of a fish, whence the name. Oolitic limestones can apparently be formed in several ways. Some of them, and especially the coarser types, are formed by the cementation of grains derived from an older limestone, and rounded by rolling in water; they are in fact sands composed of grains of limestone instead of quartz. Others, and perhaps the majority, are formed by a chemical process; water becomes saturated with calcium carbonate, which is deposited in layers round a nucleus of any foreign substance, such as a quartz-grain or bit of shell. These show under the microscope a well-developed concentric structure. A third type appears to be formed by the activity of organisms, probably some kind of algae, though this is not very well understood. When the grains are of some considerable size the rock is often called *pisolite* (pea-stone). Some well-known examples of pisolite are formed by hot springs, as at Karlsbad in Bohemia and Vichy in France.

By far the greater part of the limestones are formed either directly or indirectly through the agency of animals and plants. A very large proportion of the marine invertebrata possess a calcareous shell or skeleton of some kind and after the death

[1] Rogers and Du Toit, *Geology of Cape Colony*, 1909, pp. 401–404.

of the animal these structures accumulate in great masses on the floor of the sea, forming shell-beds and other such deposits, in water of varying but not very great depth. Most of these are loose and incoherent at first, but are afterwards cemented into solid rock by partial solution and redeposition of the calcium carbonate. Some calcareous deposits however are solid and coherent from the beginning, the best example of this class being coral-rock. Of late years it has been shown that calcareous algae play a considerable part in the building up of coral-reefs.

The floors of many shallow seas, such as the English Channel, are largely covered by layers of more or less broken shells, constituting *shell-sand,* which is a true mechanical deposit, but of organic origin. From this there is a regular transition to the shell banks that are so common a feature of shore lines and shallow water, especially where exposed to prevailing winds, as on the western coasts of the British Isles and of Holland. When partially solidified these give rise to such deposits as the Pliocene Crags of Norfolk and Suffolk, and when completely cemented and hardened they form shell-limestones, as in the Lower and Middle Jurassic (Lias, Great Oolite and other formations). Sometimes beds of oysters and other similar bivalves may be buried in sediment and preserved as shelly bands among clays, shales and other non-calcareous deposits. Many limestones of all ages have been formed from accumulations of whole and broken calcareous organisms in water of various depths, and among the older systems these animals often belong to groups now rare or even extinct. A notable example is afforded by the crinoid limestones of the Lower Carboniferous. Though one of the most abundant of all calcareous animals at that time, the crinoids are now rare and quite negligible as rock-builders. Associated with them are great numbers of corals and brachiopods, and the whole mass was probably formed in clear though not deep water. The origin of the Wenlock Limestone has given rise to some discussion; though rich in corals, it does not appear to have been a true coral-reef, but was perhaps more of the nature of a shell-bank, formed in a warm or tropical sea.

Chalk is a very special variety of limestone, of a white or greyish colour and of peculiar constitution. It consists very largely of the tests of foraminifera and fragments of shells of lamellibranchs, brachiopods and other marine calcareous organisms. The Chalk appears to have been laid down in a partially enclosed sea of considerable size, perhaps something like the Gulf of Mexico. The Chalk generally contains a good many flints in bands and nodules, but otherwise it is remarkably pure, sometimes containing 97 or 98 per cent. of calcium carbonate. Hence it is of special value as a source of lime, and is extensively quarried and burnt for that purpose. Some varieties contain a considerable proportion of phosphoric acid, and are therefore valuable as a source of this plant food, as well as of lime. The wide area occupied by the Chalk in the east and south of England makes this formation of special importance to agriculture. In the east of England, especially in Cambridgeshire, the lowest division of the Chalk is somewhat different from the rest. It contains more argillaceous matter, and in fact it forms a transition between the limestones and the marls. It is therefore known as the Chalk Marl, and it is largely used in the manufacture of Portland cement, since in it clay and carbonate of lime are already mixed in the correct proportions.

Marl. This term is often used somewhat vaguely by agricultural writers, but in the strict sense it indicates a clay with a considerable proportion of carbonate of lime, and marl is therefore intermediate between clay and limestone. Marls when newly formed are soft and plastic, but are often consolidated at a later stage into argillaceous limestones. They are very commonly formed in fresh-water lakes, and in estuaries, consisting to a large extent of broken shells and calcareous mud. A marine variety, the Chalk Marl, was described in the last section. The sites of the recently drained meres of the fenland are often occupied by deposits of *shell marl*, a soft, incoherent, white deposit, chiefly composed of the shells of fresh-water mollusca.

Dolomite-rock or magnesian limestone. Most limestones contain more or less magnesium carbonate, since this substance is found in small proportion, not usually exceeding 8 or 10 per cent., in calcareous organisms, but in some rocks the mineral dolomite is an important, and sometimes almost the only constituent. The composition and characters of the mineral dolomite have already been described (see p. 9). The origin of dolomite-rock has given rise to a good deal of controversy, and even yet the question is still unsettled. In a few instances it appears that the dolomite-rock is an original formation, having been produced directly by chemical or physical processes, such as the evaporation of solutions rich in carbonates of lime and magnesia. But this direct origin appears to be rare, and in the majority of cases it is quite certain that the dolomite-rock was at first a normal limestone, consisting mainly of calcium carbonate, with the usual small admixture of magnesia and other impurities. The conversion of ordinary limestone into magnesian limestone is known as dolomitization, and it has taken place to some extent in many or most of the older calcareous sediments. It appears to be specially liable to occur in coral-rock, and this fact is of much importance in connexion with the origin of the change. The hard parts of corals consist of aragonite, a form of calcium carbonate less stable than calcite, and therefore more liable to take part in chemical reactions.

Sea water is notably rich in magnesium salts, and it is generally believed that dolomite is formed by a reaction between calcium carbonate and these magnesium salts, especially magnesium sulphate, resulting in the formation of dolomite and calcium sulphate. This theory is supported by the common association of the mineral gypsum with beds of dolomite-rock. The reaction is apparently favoured by high temperature and a pressure of from one to five atmospheres. Such conditions are attained in tropical regions from the surface of the sea down to a depth of a little over 20 fathoms, and in such situations it appears from observations on coral-reefs that dolomitization is most active. In this way were probably formed the famous "Dolomites" of the Tyrol, which are of Triassic age. The most extensive mass of dolomite-rock in Britain is that known as the Magnesian Limestone, of Permian age. This forms a broad band extending from Nottingham to the mouth of the

Tyne, and reaching a maximum thickness of 800 feet in Durham.

Many of the older limestones are found to be largely dolomitic near the upper surface and along the major joints, but unaltered in the inner and lower parts. It is often easy to show that such masses have been overlain by Triassic deposits containing rock-salt and gypsum, and it is clear in many instances that the dolomitization was effected by percolation of salt solutions from the overlying rocks.

It is only rarely that dolomitization is complete; usually the rock contains an excess of calcium carbonate and therefore consists of a mixture of calcite and dolomite. In the following table are given analyses of some typical magnesian limestones, British and foreign:

	Port Shepstone, Natal (Hatch)	Langkofl, Tyrol (Skeats)	Newcastle, (Rosenbusch)
SiO₂ etc., insol. ...	4·29	0·02	5·00
FeO	0·47⎱	—	1·50
Al₂O₃	0·03⎰		
CaCO₃	57·75	54·70	57·50
MgCO₃	37·46	45·30	35·33
	100·00	100·02	99·33

Dolomite-rock, when burnt in a kiln like limestone, is converted into a mixture of lime and magnesia. Lime made from a highly magnesian limestone however is not generally suitable for agricultural purposes, as it is more caustic than pure lime, and has a "burning" effect on crops. Magnesian limestone is however employed to a considerable extent in certain metallurgical processes.

Ironstone. The ironstones constitute a large group of rocks, almost exclusively of sedimentary origin, and formed in various ways, but all agreeing in containing a sufficiently large proportion of iron to make them of value as ores of that metal. Iron ores of purely igneous origin are rare and need not be further considered here. The sedimentary ironstones may be divided roughly into two classes; those which were ironstones from the beginning of their existence, and those which were produced by alteration of limestones.

The formation of ferruginous deposits can be seen in progress at the present time in certain lakes and swamps, where there is much decaying vegetation. All natural waters contain iron in solution, chiefly as salts of various organic acids of the humic group. Under certain conditions this iron undergoes oxidation, forming insoluble hydrated oxides, of which limonite, $2Fe_2O_3 . 3H_2O$, is the most common. This oxidation is probably brought about by the action of bacteria[1], or according to other authors by diatoms. Perhaps the most striking example of such a method of iron ore formation is afforded by the "lake-ores" of Sweden, but a similar process is in operation in many swamps and marshes in other countries. Its chief agricultural importance lies in the fact that in some wet peaty soils a hard layer of hydrated iron oxide is formed a few inches below the surface and this is often thick and hard enough to interfere seriously with proper cultivation. This is one form of "pan" (see also p. 140).

In the Coal Measures are found abundantly nodules and beds of impure earthy ironstone (clay-ironstone and black-band ironstone). These varieties consist of carbonate and oxides of iron in varying proportions, together with silica, alumina, phosphoric acid and other impurities, and often a good deal of carbonaceous material is present, derived from the decay of organic matter of animal and vegetable origin. They are sometimes very fossiliferous. The origin of these ironstones is obviously similar to that of the modern lake-ores and bog-ores.

The mineral haematite, Fe_2O_3, one of the most valuable ores of iron, occurs in mineral veins, and more commonly in large masses in fissures and cavities in sedimentary rocks, especially in limestones. Perhaps the best-known occurrence is in the Furness district of Lancashire, where it gives rise to the great iron industry of Barrow-in-Furness. Further north, in west Cumberland, haematite is also found in masses in the Skiddaw Slates, as well as in the Carboniferous. It appears to have been deposited mainly by water holding it in solution percolating through from overlying ferruginous strata.

---

[1] Winogradsky, "Über Eisenbakterien," *Botanische Zeitung*, 1888, p. 260.

Some of the most important ironstones in this country are those that have been produced by alteration of limestones. This change is in its simplest form a replacement of calcium carbonate (calcite or aragonite) by ferrous carbonate (chalybite). The change is brought about by solutions containing iron salts, such as are found percolating through all rocks. Since aragonite is less stable than calcite, this mineral, if present, is attacked first. The iron carbonate first formed is, however, itself unstable and easily undergoes further oxidation to limonite, haematite or even in rare cases to magnetite. Ironstones formed in this way are often highly fossiliferous, and show all the structures characteristic of limestones, especially the oolitic structure. The best examples in this country are the ironstones of the Jurassic system in the Midlands and in Yorkshire. The Cleveland Main Seam, in the Middle Lias, is as much as 25 feet thick at Eston, near Middlesborough, and has given rise to an enormous mining industry in the Cleveland area. The Top Seam or Dogger of the same district is of much inferior quality, but extends over a very large area. At about the same horizon (the base of the Inferior Oolite) are the well-known ironstones of the Northampton Sands series of Lincoln, Rutland and Northampton. Ironstone is also worked in the Lower Lias at Frodingham and Scunthorpe in Lincolnshire. In times past a great iron-smelting industry also existed in the Weald of Sussex and Kent, where wood was used for fuel, but on the failure of the wood supply the workings were abandoned.

The presence of beds of ironstone is not directly of much agricultural importance, but indirectly it has one useful result. Ironstones are often comparatively rich in phosphoric acid, and weathered material distributed over the soil by rain-wash from an outcrop at a higher level may be a useful natural source of this constituent. Such is found to be the case with some of the soils in the neighbourhood of the Northampton ironstones. Again perhaps the cheapest and most convenient of all "artificial" phosphatic manures is basic slag, which is a by-product in the manufacture of steel from phosphatic iron-stone.

In many countries, though not in Britain, parts of the

crystalline schists of the Archaean systems are highly ferruginous, often being very rich in haematite or magnetite. This appears to be due largely to secondary enrichment by solutions either before or during metamorphism. Good instances are afforded by the Penokee iron-bearing schists of Michigan, and the quartz-haematite and quartz-magnetite schists of the Swaziland system in South Africa. The latter are curious rocks, showing conspicuous bands of white and black or red (the "calico-rock" of the miners and prospectors). Although rich in iron, such rocks are generally too siliceous to be worked profitably as ores.

Silicification of limestones. Flint and chert. It has often been observed that calcium carbonate and silica are able to replace each other in rocks, the process going on sometimes in one direction, sometimes in the other. Most commonly however calcium carbonate is replaced by silica, and the rearrangement of the silica originally in the sediment as well as addition of this substance from outside, gives rise to some very well-known rock-types; of these flint and chert may be specially mentioned.

The simplest case of all is where the aragonite or calcite of a limestone has been replaced molecule for molecule by silica, preserving perfectly all the structures of the limestone, organic and otherwise, these now consisting of chalcedony instead of the original mineral. A well-known example is to be found in the cherts of the Portland series, which often show very perfectly preserved masses of reef-corals. In this case the silica must have been brought in from outside. More commonly however chert is formed by rearrangement of silica already existing in the sediment, very often in the form of sponge remains. The skeletons of most sponges consist of colloidal silica, which is more readily soluble than the crystalline form. This is dissolved by percolating water and often deposited again almost immediately in the solid state, sometimes in a concretionary form, around some originally siliceous nucleus; sometimes in a more than usually siliceous layer or bed. It appears that silica tends to travel towards points of more than the average concentration and any unusually siliceous patch will therefore act as a centre for deposition. Some of the most highly siliceous bands of the Lower and Upper Greensand seem to have been largely composed of remains of sponges, and these originally very siliceous beds underwent secondary enrichment, forming well-marked layers of chert. The same remark applies to the cherts of the Carboniferous Limestone in west Yorkshire and North Wales; these are of some commercial value in the manufacture of certain kinds of pottery.

The best-known siliceous concretions are the *flints*, found originally in the Chalk, but now forming such an important constituent of nearly all the later formations of the south and east of England. Owing to their stability and great hardness flints are passed on from one formation to another with little change, except a slight reduction in size by mechanical wear. Flints vary much in colour, from black, though every shade of grey, brown and yellow to nearly white; when freshly dug out of the Chalk they are generally covered with a thin white film, but this is soon lost on weathering or rubbing. The variations in size and shape are endless, but generally quite irregular and more or less nodular. Flints may be described briefly as concretionary masses of cryptocrystalline (i.e. very finely crystalline) silica, formed by concentration of silica, formerly more or less disseminated through the Chalk, around certain points or along certain planes. The core of a flint is often a sponge or other fossil, and the so-called tabular flints may be formed either along bedding-planes or along joint-planes inclined to the bedding. From a study of the microscopic characters of Chalk it appears that the silica of the flints is obtained from sponges, radiolaria, diatoms and other organisms originally forming part of the Chalk itself.

There appears to be no essential difference between flint and chert; it is simply a matter of nomenclature, the fact being that the characteristic siliceous bodies from the Chalk are universally called flints, while siliceous masses from all other formations, whether of similar or different origin, are called chert. It is hardly necessary to point out here the important part played by flint as a material for tools and weapons in the early stages of human development. This subject, which is of the greatest interest, belongs to pre-historic archaeology rather than to geology. The farmer is concerned only with flints as constituents of soils and of superficial deposits in general, and sometimes as road-material; for this latter purpose they are very badly suited owing to their hardness and sharpness; they are specially destructive to rubber tyres.

Phosphatization and phosphatic deposits. The great agricultural importance of phosphoric acid as a constituent of soils and of manures and as a plant food necessitates a somewhat full account of the origin and characters of naturally occurring phosphatic deposits.

The ultimate source of phosphorus compounds in sediments and in soils is to be sought in the mineral apatite, a widely distributed constituent of igneous rocks of all kinds and of all ages. In a few places, as for example in south-western Norway, apatite occurs in workable quantities in veins and segregations in igneous rock (gabbro or dolerite). In other parts of Norway, in Spain and in Canada great masses of crystallized apatite are found; these are believed to have been originally of sedimentary origin and afterwards crystallized by metamorphism. All of these are now largely employed in the manufacture of superphosphate. For all practical purposes apatite may be regarded as pure calcium phosphate; the small amount of fluorine or chlorine present in the mineral is of no agricultural significance; consequently superphosphate made from apatite is generally of very high grade.

When a rock containing apatite is weathered the apatite appears to dissolve, as it is very rarely found in the form of crystals in detrital sediments. Thus the calcium phosphate gets into circulation in natural waters, and sets up an endless cycle of chemical and physiological processes, passing into the sediments and soils, from thence into plants and ultimately into the bones and tissues of animals. The excreta of animals and the decay of their bodies after death continue the process to further stages, and the whole story is evidently of very great complexity. The main point is that natural waters always contain soluble compounds of phosphoric acid. These are in part taken up directly by plants and animals, the other part promoting chemical changes in rocks and soils. It is clear that when such phosphatic solutions come in contact with carbonates, the latter are decomposed and phosphoric acid is substituted for the carbonic acid, forming insoluble phosphates.

The study of deposits of guano and of the rocks underlying them has thrown much light on this subject. Guano is very rich in soluble phosphate, probably existing to a large extent as ammonium phosphate and this is often carried down into the rocks below. When, as often happens, the guano lies on coral-rock, it is often found that the calcium carbonate is more or less completely changed to calcium phosphate. It is believed that the great phosphate deposits of Christmas Island, in the Indian Ocean, were formed by this means, although the guano has now completely disappeared. Cases are also known where igneous rocks have been converted into phosphates of iron and alumina by a similar process.

Where deposits of guano have been leached out by rain, most of the nitrogenous matter, being easily soluble, is removed, and the less soluble phosphates are left behind. Furthermore, since calcium phosphate is more soluble than phosphates of iron and aluminium, excessive washing will remove the former and leave the latter behind. Hence some naturally washed guano residues, where the leaching process has not gone too far, are good phosphatic manures in their original state, but if little but iron and alumina compounds are left the manurial value is small. In an analysis of a natural phosphate it is insufficient and misleading to determine only the total phosphoric acid; it should also be stated what proportion of this is present as calcium phosphate, since this alone is of much value.

All the larger natural deposits of calcium phosphate, or phosphorite, are probably of marine origin and they are apparently always due directly or indirectly to organic agencies. Unlike apatite, they are amorphous and may be either compact, earthy or concretionary. Nodular forms are very common in rocks of many ages, and have been dredged up from considerable depths among the deposits of the modern sea-floor, nearly always in association with the mineral glauconite. They must therefore be regarded as, at any rate in part, original constituents of sediments. But it is clear that many phosphatic nodules were originally calcareous and only became phosphatic at a later stage. Such were the co-called "coprolites[1]" of the Cretaceous strata of the south and east of England, and many others of similar origin. The same statement is true of many phosphatic beds, layers and masses in rocks of various ages.

In many pebble-beds of shallow-water marine origin, and especially in those associated with local unconformities and "wash-outs" there are to be found phosphatic nodules of a special character, nearly always accompanied by glauconite. Many of the phosphatic nodules (*coprolites* of the trade) are obviously rolled fossils, or fragments of fossils, derived from older rocks during the concurrent denudation; besides these there are often many concretionary masses of phosphate of the

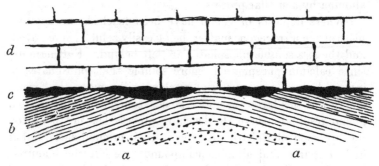

Fig. 30. Mode of occurrence of the Cambridge Greensand.
a, Lower Greensand; b, Gault; c, Cambridge Greensand; d, Chalk.

type before described. The fossils must have been originally calcareous and they have been phosphatized while lying uncovered on the sea-floor. Shallow seas, especially where wave-action and currents prevail, always abound in animal life, and the phosphoric acid is derived from the excreta and from the decaying remains of these animals, among which fish, and in past ages gigantic reptiles, are the most prominent. The best example of this process is afforded by the well-known Cambridge Greensand, which 30 or 40 years ago gave rise to an important

---

[1] The term *coprolite* has been much misapplied. Originally it meant the fossil excreta of animals, but it was later used as a commercial name for any kind of phosphatic nodules, of whatever origin.

industry. Owing to a slight local disturbance a small unconformity was produced between the Gault and the Chalk; most of
the finer material of the Upper Gault was washed away, the
pebbles, nodules and fossils remaining behind to be phosphatized.
Many of the fossils of the Gault were already more or less
phosphatic and underwent a secondary enrichment, so that their
percentage of phosphoric acid is now very high. The animals
which were living at the time and were entombed in the Greensand also underwent a lower grade of phosphatization. These
can be distinguished by their paler colour. Of somewhat
similar origin and character are the phosphatic nodule beds of
the Lower Greensand, formerly worked at various localities in
Bedfordshire and Cambridgeshire. At various horizons in the
Jurassic and Cretaceous clay-formations there are layers of
more or less phosphatic nodules, and these have undoubtedly
in part been the source of the concretionary nodules of the
"derived" pebble beds just described; with them may be
compared the layers of phosphatic concretions in the Ecca
shale near Ladysmith. These are associated with fish-remains,
which were no doubt the source of the phosphoric acid.

The frequent association of phosphates and glauconite has
already been mentioned, and this is a point of some importance;
glauconite is a conspicuous mineral, easy to detect owing to its
green colour, and its presence affords a useful indication of the
possible existence in the same deposit of useful supplies of
phosphate.

Calcium phosphate is a constituent of some varieties of
Chalk, and its presence is of considerable importance, since lime
made from phosphatic Chalk obviously serves a double purpose,
enriching the soil in phosphoric acid as well as in lime. In
some varieties the phosphate appears to be finely disseminated
throughout the rock, while in others it occurs as white, grey or
yellowish nodules, apparently of concretionary origin, as in
parts of the Lower Chalk of Cambridgeshire. Phosphatic
Chalk has been worked commercially on a considerable scale at
various places in the north of France, in the departments of the
Somme, Oise and Nord, and especially at Ciply in Belgium.
The workings at the latter place yielded 85,000 tons of phosphate

in 1884. The Ciply Chalk is greyish-brown in colour, and the phosphate exists in the form of minute brown grains, all of which can be referred to an organic origin, many being casts of Foraminifera. A bed of phosphatic Chalk of very similar character exists at Taplow. Here also the phosphate is in the form of brown grains and fragments, many of these being evidently bits of scales and teeth of fish, as well as Foraminifera and shell fragments, the latter having evidently undergone secondary phosphatization, as in the Cambridge Greensand. Some oval pellets appear to be excreta of fish and doubtless much of the phosphate is derived from this source.

The following analyses, somewhat condensed from the original publication[1], show the composition of samples of phosphatic Chalk from Ciply and from Taplow.

	Ciply	Taplow
Organic matter and moisture ...	2·83	3·0
Lime	53·24	53·7
Iron oxide and alumina	1·01	·9
Potash and soda	·19	·3
Carbon dioxide	28·10	28·7
Phosphoric acid	11·66	11·6
Silica	1·96	·5

The similarity in composition is here very striking. The Ciply phosphate is concentrated by mechanical processes or by washing, and the natural weathered product found in pockets in the rock contains up to 67 per cent. of calcium phosphate, the more soluble carbonate having been removed by percolating water. This is obviously a valuable fertilizer.

Salt deposits. These constitute a peculiar and well-marked class of aqueous accumulations, some of which are of the very highest agricultural importance; as examples may be mentioned rock-salt, kainite and nitrate of soda. Although for the most part of purely chemical origin, they are as a matter of practical convenience classed with the sediments by nearly all authors. All the substances here treated together agree in the fact that they are easily soluble in water, and therefore their accumulation in the solid form can only take place under special conditions,

---

[1] Strahan, "On a phosphatic Chalk...at Taplow," *Quart. Journ. Geol. Soc.* vol. XLVII. 1891, p. 356, and vol. LII. 1896, p. 463.

the most important of these conditions being absence of rainfall. Hence salt-deposits are specially characteristic of arid regions. It is only in the desert zone that the requisite climatic conditions are attained.

All natural waters contain more or less dissolved mineral matter, and in fact the distinction between fresh and salt water is merely one of degree; the water is called salt when it contains enough dissolved matter to impart a distinct taste; when tasteless it is called fresh. Now when a salt solution is evaporated the water is driven off in the form of vapour, and the salt remains behind. A salt solution, however dilute at first, can be concentrated by evaporation, and fresh water thus becomes salt. Such a process occurring naturally on a large scale has given rise to most of the salt-deposits known to us, the chief exception being in those instances where masses of sea-water may become isolated and dry up.

The process last mentioned sometimes occurs on a comparatively small scale in lagoons along the sea-coast; owing to slight earth-movement, or to growth of a shingle bar, or some similar cause, communication with the sea is cut off, and if the climate is warm enough the water will dry up, leaving the salts behind. An artificial adaptation of this process is employed in some warm regions for the manufacture of salt, as for example in the salt pans on the Mediterranean coast of France. On a very large scale the formation of certain inland seas, such as the Caspian, is an exactly analogous process. At one time the area now occupied by the Caspian, the Sea of Aral and other salt lakes of south-western Asia was in free and open communication with the Arctic Ocean. As a result of earth movements this communication was cut off, and the whole area converted into a self-contained closed drainage basin with no outlet to the sea. A little consideration will show that the maintenance of such a condition must depend upon the ratio between loss and gain of water; that is to say, if the rainfall in the drainage area exceeds the loss by evaporation the basin will gradually fill up with water till an outlet is established at the lowest point on the rim; in course of time all the salt will be washed out and the lake will become fresh. On the other hand, if evaporation is in excess the lake will become salter and salter, till the water is saturated and salts are deposited.

But a salt lake may also originate without any connexion with the sea at any period of its existence. If by some means a fresh-water lake loses its outlet, it will gradually become salt on account of the dissolved mineral matter brought in by rivers, and the waters of the lake will eventually reach saturation point. Lakes without outlet are common in arid regions, in fact wherever evaporation exceeds rainfall, and it is in this way that most of the well-known salt lakes have been formed, for example, the Dead Sea, and the Great Salt Lake of Utah. Both are relics of once much larger fresh-water lakes, which have nearly dried up, and the formation of the Dead Sea was specially favoured by the fact that the rocks of the Jordan valley were originally very rich in salt. This is also a region of little rainfall and very high temperature.

When a mixed salt-solution, such as sea-water, is evaporated, the salts are deposited in a certain definite order, depending on the relative solubilities of the salts present. The first salt formed from natural waters is calcium sulphate, which is deposited as gypsum or anhydrite according to temperature, the former being the more common. This is followed by sodium chloride (common salt) while the salts of potassium and magnesium, being much more soluble, remain in solution till nearly all the water is evaporated, and it is only in very rare cases that they have been deposited in nature. The laws governing the crystallization of potassium and magnesium salts are very complex, and a large number of minerals have been formed by this process, the most important from an agricultural point of view being kainite and carnallite.

The best example of what is known as a "complete" salt deposit (that is, one including potassium and magnesium salts) is found at Stassfurt in Germany, and this is now the chief source of artificial potash manures.

The chief salts found at Stassfurt are enumerated in the following table:

Rock-salt	...	NaCl.
Sylvine...	...	KCl.
Carnallite	...	$KCl . MgCl_2 . 6H_2O.$
Polyhalite	...	$K_2SO_4 . MgSO_4 . 2CaSO_4 . 2H_2O.$
Kainite	...	$KCl . MgSO_4 . 3H_2O.$
Gypsum	...	$CaSO_4 . 2H_2O.$
Anhydrite	...	$CaSO_4.$

In the lowest part of the shaft is about 700 feet of rock-salt; then comes the polyhalite region, about 200 feet thick. This is succeeded by about 200 feet of magnesium sulphate with some carnallite, while the highest layer, about 100 feet thick, contains kainite, carnallite and magnesium chloride. Resting on the salt-deposit is a bed of impervious clay, and to this is doubtless due the preservation of all these highly soluble minerals. Kainite, carnallite and other potash salts are also found at Kalusz and Aussee in Austria.

All the evidence indicates that these salts were formed by the evaporation of enormous volumes of sea-water, although it is not quite clear how the actual evaporation was brought about; whether in a closed basin or in an area still in communication with the open sea. Space will not permit of a full discussion of this interesting subject[1].

Deposits of rock-salt and gypsum, without salts of potassium and magnesium, are found among sediments of both ancient and modern date in many parts of the world; in fact they are highly characteristic of the desert facies of deposition, and have evidently been formed for the most part in salt lakes without any connexion with the sea.

As an example we may take the Trias salt-beds of the British Isles; these are found in Cheshire, Worcestershire, near Middlesborough in north-east Yorkshire, and in the neighbourhood of Belfast (Carrickfergus). The salt and gypsum are interstratified with beds of fine-grained red marls and red and yellow sandstones, the latter often showing the conspicuously rounded grains that are so characteristic of desert sands (millet-seed sands). The so-called marls are not generally real marls, that is, they are not markedly calcareous, but are simply

[1] For further details and references see Hatch and Rastall, *The Petrology of the Sedimentary Rocks*, London, 1913, pp. 93–108.

sediments of unusually fine texture; they were formed by dust drifting before the wind and accumulating in salt lakes, in which the gypsum and rock-salt were also deposited by concentration of the saline water.

The greatest known thickness of rock-salt was found in a deep boring at Sperenberg, near Berlin; this passed through about 4000 feet of salt without reaching the base. Very thick deposits are also known at Wieliczka in Austrian Poland, and at Parajd in Transylvania; these are of Tertiary age.

Modern salt lakes can be conveniently divided on chemical grounds into three groups, as follows:

(*a*) Salt lakes proper, such as the Dead Sea and the Great Salt Lake of Utah; these contain chiefly chlorides of sodium and magnesium; they are saline to the taste, but not bitter.

(*b*) Bitter lakes; these owe their peculiar taste to the abundance of sulphates, especially sodium sulphate (Glauber salt) and magnesium sulphate (Epsom salts). Some well-known bitter lakes on the Isthmus of Suez were destroyed by the construction of the Suez canal. They also occur in the Aralo-Caspian basin and in western America.

(*c*) Soda-lakes or alkali lakes; these contain abundance of sodium carbonate and sodium bicarbonate. The natron lakes of the Egyptian desert are of this nature and others are found in the Great Basin of western America.

Nitrates. Nitrates are formed in all fertile soils by the action of certain specific bacteria on nitrogenous organic matter, and a small amount of nitric acid is also brought down from the atmosphere by rain. Nitrates are highly soluble and are either taken up at once by plants or carried away in the drainage water. It is but rarely and only under special circumstances that nitrates can accumulate in the solid form. Nitrate of lime is found in considerable quantity in certain caves in Kentucky and Indiana, and it is believed to be formed by oxidation of guano. Potassium nitrate (saltpetre) has been observed as an efflorescence on the surface of the soil in certain dry regions in north-western India. But the chief commercial and agricultural substance of this class is nitrate of soda or Chile saltpetre, which is found in great masses in certain parts

of South America. It has also been noticed in association with
boron minerals in southern California. The nitrate beds of
South America are found in the rainless district of Peru and
Chile, especially in the deserts of Atacama and Tarapacá,
these being perhaps the driest regions of the world. They are
often at a considerable distance from the coast, even as much
as 180 miles, and at considerable elevations, up to 5000 feet
above sea-level. These facts seem to militate against a marine
origin, as will be mentioned later.

The crude sodium nitrate, called *caliche*, is interstratified
with sediments of various kinds in association with beds of
common salt and borax; the following is a section of a typical
"calichera":

Sand and gravel	2 inches.
Gypsum	6 ,,
Compact earth and stones	10 feet.
Caliche	5 ,,
Clay below.	

Sometimes the bed of caliche is as much as 12 feet thick and
the details vary slightly in different localities.

The caliche is by no means pure and it varies widely in
composition; the chief impurities are sodium chloride, sodium
sulphate, calcium sulphate and magnesium sulphate, together
with a small proportion of borates, iodates and ammonia
compounds. Some samples of caliche contain a little potassium
perchlorate, a salt which is supposed to be very injurious to
vegetation; such samples should be avoided.

Many suggestions have been made as to the origin of the
nitrate deposits; the following are some of the less improbable
theories. Ochsenius[1] supposed that the basis of the whole
process was to be found in beds of rock-salt existing in the
Andes; these were dissolved by rain and carried down to lower
levels, where the solution was acted on by carbon dioxide of
volcanic origin, forming sodium carbonate. This is then decomposed by dust containing nitrogen and ammonia carried by
the wind from beds of guano on the coast, forming nitric acid
and eventually sodium nitrate. This cause seems inadequate

---

[1] Ochsenius, *Die Bildung des Natronsalpeters,* Stuttgart, 1887.

to produce the observed results. Another somewhat similar theory[1] supposes that beds of guano have undergone bacterial decomposition, forming nitric acid, and then calcium nitrate, which reacted with sodium sulphate in the underlying strata, forming sodium nitrate and gypsum, thus:

$$Ca(NO_3)_2 + Na_2SO_4 = CaSO_4 + 2NaNO_3.$$

It is certainly a fact that gypsum is found along with the caliche, but the absence of phosphorus compounds seems to disprove an origin from guano, in which phosphates are always present in large quantity.

An older theory is that of Noellner, who referred the origin of the nitrate to the decomposition of great masses of seaweed, forming nitric acid, the rest of the process being as last described. This accounts satisfactorily for the presence of iodine and the absence of phosphate, but the chief objection is the great elevation and distance from the sea at which the deposits are now found, implying an enormous rise of the land in recent times. It has long been known, especially from Darwin's observations, that the west coast of South America is undergoing elevation even now, but the movements postulated seem too great to be credible.

Owing to the association of boron salts with nitrate of soda, both in Chile and in California it has been suggested that these deposits are ultimately of volcanic origin. This suggestion is plausible, but in the present state of knowledge not fully established. We must be content with the statement that the origin of nitrate of soda is as yet an unsolved problem.

---

[1] Plagemann, *Die Dungstoffindustrie der Welt*, Berlin, 1904.

CHAPTER V

SUPERFICIAL DEPOSITS

Of the highest importance from the agricultural point of view are the superficial accumulations of varying origin which in so many places mask the solid rocks, and by their weathering and alteration give rise to the soil and subsoil. In this chapter the actual cultivated layer of the soil is not dealt with, and attention is confined to the origin and characters of the deposits of recent origin and very varying thickness that constitute the subsoil in many regions. These, when present, obviously have a controlling influence in determining the agricultural value of the soil, and their study is of the utmost importance to the farmer.

For these deposits many classifications have been proposed, but none are wholly satisfactory, partly owing to the infinite variety of the deposits, and partly because apparently similar accumulations may have been formed in very different ways. The classification here adopted is based upon that proposed by Merrill[1], but certain modifications have been introduced, in order to render the treatment more suitable for the special purpose of this book, and the original American terminology has been in some instances replaced by terms better understood in England.

The superficial deposits may be divided primarily into two broad groups, namely, *sedentary* and *transported*. The sedentary deposits are formed mainly by the weathering and alteration of the underlying rocks, and their character is determined entirely by the nature of these. Some also are

[1] Merrill, *A Treatise on Rocks, Rock-weathering and Soils*, New York, 1897, p. 300.

formed wholly or in part by the growth and decay of vegetation. Transported deposits on the other hand have been formed of material brought from a greater or less distance by the various geological agents: gravity, water, ice and wind. They need show no resemblance to the underlying rocks and generally differ markedly from them. It is evident that transitional forms may exist, partly sedentary and partly transported, and many soils and subsoils are thus of mixed origin.

The character of both sedentary and transported deposits is to a large extent controlled by climate, but it is necessary to remember that in some instances the dominant factor is the climate of a period which has long passed away, as in the well-known instance of the glacial deposits of Europe and North America. The special influence of climate on rock-weathering has already been dealt with (see Chapter II), and this is evidently of fundamental importance also in connexion with the sedentary deposits as above defined. Many of them are in fact merely masses of weathered rock, that have suffered disintegration, but have not been removed by the agents of transport.

Sedentary deposits. These are again subdivided into two main groups:

> (a) Residual deposits,
> (b) Cumulose deposits.

The term residual signifies material that has been left behind during the operation of the ordinary geological agents of weathering and transport, and from this statement the general character of the material is almost self-evident. Cumulose deposits on the other hand are mainly formed by the growth and decay of plants and to a much less extent of animals.

(a) *Residual deposits.* Since the majority of rocks are not homogeneous, but are made up of materials of varying chemical and physical properties, the effects of weathering on these different constituents are also variable. Some minerals are more readily soluble than others, and are more easily attacked by acids or other agents. Hence there is always a tendency for the more stable portions of the rock to remain behind, while the less stable are removed. A simple example is afforded by certain rocks composed of grains of quartz and other resistant minerals embedded in a cement of calcite (calcareous sandstone). The effect of weathering on such a rock is to remove the more soluble and softer cement, leaving a residue of sand. Precisely similar are the results of weathering of sandstones with a cement of iron oxide; this is easily dissolved by water containing organic acids; hence sandstones in regions where there is little mechanical removal of solid material are covered by a thick layer of residual sand. Again, conglomerates consisting of hard pebbles with a softer or more soluble matrix yield gravels. Such residual gravels formed by the weathering of conglomerates are remarkably well developed on the Witwatersrand and in other parts of the southern Transvaal, where owing to the absence of transport, weathering may extend to a depth of several hundred feet[1]. Of very similar origin are the plateau gravels and cannon-shot gravels of Cambridgeshire and Norfolk, formed from glacial deposits, especially boulder-clay, by the removal of the finer clayey material, leaving the larger constituents behind as a residue. These consist for the most part of flints, although far-travelled rocks are present in some quantity. One of the most remarkable examples of this process is afforded by the "Sarsen stones" of certain parts of the Downs of the south of England, especially in Wiltshire. Stonehenge and other prehistoric monuments are largely built of them. Sarsen stones are blocks of a kind of quartzite of lower Tertiary (Eocene) age, that have been weathered out *in situ*, the rest of the original Tertiary stratum having disappeared. They are closely related to the well-known "Hertfordshire Pudding Stone," a conglomerate of flint pebbles with a cement of crystalline quartz.

The Chalk of southern and eastern England is frequently covered by a layer of flint-gravel, or by a mixture of flints and clay. In some instances the flint gravels are to be regarded as simply the insoluble residue of the Chalk, the calcium carbonate having been removed in solution. In other cases, however, the gravels are of fluviatile or glacial origin, coming therefore

---

[1] Hatch, "Conglomerates of the Witwatersrand," in *Types of Ore Deposits*, 1911, p. 202.

within the category of transported deposits. The origin of the deposit known as the clay-with-flints of southern England has given rise to considerable controversy[1].

The clay-with-flints covers a considerable area in the south-eastern counties, and is of special interest in that it forms a considerable part of the subsoil of the famous experimental farm at Rothamsted in Hertfordshire. It consists of a somewhat variable mixture of flints and red clay or loam, the latter being generally somewhat stiff and pasty in character. The thickness is far from being constant and the clay sometimes descends into deep pipes in the underlying Chalk. It was formerly believed that the whole accumulation, including the clay, was simply the insoluble residue of the Chalk, since many limestones are known to yield a red clay on weathering (see p. 115). But it has been shown that the clay is present in too great quantity in proportion to the flints to have been formed solely in this way, and it is now believed that much of the material is the residue of formerly existing Eocene strata, now mainly removed. It is even suggested that, north of the Thames, the clay-with-flints may be partly or mainly of glacial origin[2]. However, south of the Thames, whether wholly Cretaceous or partly also Eocene, it is at any rate a true residual deposit.

The greatest developments of residual weathering products are found in tropical regions, where chemical and bacterial processes are specially active, and thick vegetation prevents transport of the disintegrated material. Under these conditions weathering may extend downwards for hundreds of feet and there is a perfectly gradual transition upwards from unaltered rock to soil. This helps to account for the extreme fertility of some tropical districts. Under certain well-defined climatic conditions, however, disintegration is not by any means uniform throughout the mass of the rock; on the contrary, in countries subjected to great daily variations of temperature, rocks, especially igneous rocks, tend to break up into masses along their dominant joints. The edges and corners of the blocks are

[1] Jukes-Browne, *Quart. Journ. Geol. Soc.* vol. LXII. 1906, p, 132.
[2] Sherlock and Noble, *Quart. Journ. Geol. Soc.* vol. LXVIII. 1912, p. 202.

rounded off by weathering and the ground is covered with piles of blocks of varying size and shape. For these the name of *boulders of disintegration* has been suggested. Well-known examples are the granite kopjes of Mashonaland (e.g. the Matoppo Hills) and the dolerite kopjes of the Karroo in South Africa. Very similar in their origin are the masses of rock debris that cover the outcrops of granite and other igneous rocks in many parts of the world. In our own country masses of rock fragments of this nature on a large scale are mainly formed at high elevations or on very steep slopes and are therefore not of much agricultural importance. In some districts however, where climatic conditions are suitable, the outcrops of large masses of granite are so thickly covered with an accumulation of partly weathered blocks, that even at low elevations and on comparatively level surfaces the ground is rendered perfectly useless. Such for example is the case in certain parts of Ireland, especially in Donegal.

Perhaps the simplest of all residual deposits are those formed by solution of part of the rock. This occurs most commonly in limestone and dolomite-rock. Calcium and magnesium carbonates are soluble in water with comparative ease and the impurities present in the rock are left behind on the surface. These are mainly argillaceous or ferruginous in character and often give rise to red or brown soils. Even the Chalk is sometimes covered by a brown loamy soil of this nature, as also are the Carboniferous Limestone and, even more markedly, the Jurassic limestones of the south-western and midland counties. These limestones are often distinctly ferruginous and the iron becomes concentrated in the less soluble residue. Of similar origin is the residual deposit known as *terra rossa*, which covers great areas in southern and south-eastern Europe, being characteristically developed in what is known as the Karst area, to the east of the Adriatic. The accumulation of terra rossa is favoured by a rather deficient rainfall, so that the material is not removed by mechanical transport.

In tropical and semi-tropical regions vast areas are covered by the deposit known as laterite. This varies greatly in composition, but may be described in general terms as a mixture in differing proportions of hydroxides of iron and alumina, with, in addition, more or less manganese and titanium oxides. The latter are usually not of much importance, but some Indian laterites are of value as ores of manganese. Laterite varies in composition from nearly pure iron hydroxide to nearly pure aluminium hydroxide (bauxite).

The origin of laterite has given rise to much controversy. It may however be ascribed to a peculiar type of weathering of rocks rich in alumina and iron under certain special climatic conditions. The most characteristic feature of this type of weathering is removal of silica in solution, probably owing to hydrolysis of silicates. Although laterite is most common in tropical and subtropical countries it is not exclusively confined to them, but it appears that strongly contrasted wet and dry seasons are essential. Some authors attribute great importance to deposition of dissolved material from ascending solutions, the ascent being conditioned by rapid evaporation during dry, hot seasons. Others regard the process as mainly one of decomposition *in situ*, while it has even been suggested that the decomposition may be due to bacteria, possibly allied to the well-known nitrifying bacteria of soils. This is however mainly speculative[1].

Laterite is most highly developed on basic igneous rocks, such as the Deccan basalts of India, but it has also been formed from schist, gneiss, slate, sandstone and granite. Indian geologists recognize two types of laterite, high-level and low-level, the former being a true residual deposit, while the latter has undergone transport, occurring mostly as a cementing material in sands and gravels.

Normal laterite is a distinctly red or brown substance, often with ferruginous nodules and concretions. It is fairly soft when freshly dug, so that it can be made into bricks, but it becomes very hard on exposure to the air. Besides its very widespread occurrence in India it is also known in the Malay Peninsula, Java, Sumatra, South America and many parts of Africa; in fact it seems to be one of the commonest residual

---

[1] For a summary of recent views on the origin of laterite see Hatch and Rastall, *The Petrology of the Sedimentary Rocks*, London, 1913, pp. 326–331.

products of igneous rocks under suitable conditions of climate,
and according to Ramann it is the characteristic superficial
deposit of the tropics (see also p. 57).

In arid regions there is always a considerable upward
movement of water by capillarity from the deeper layers of
the subsoil and from the underlying rocks, to replace that lost
by evaporation at the surface. This water contains a good deal
of matter in solution, especially silica or calcium carbonate.
These substances tend to deposit themselves in the interstitial
spaces of the surface deposits, gravels, sands or soils as the
case may be, thus often producing a hard layer, either actually
on the surface or just below it. Where this occurs it offers an
insuperable obstacle to cultivation.

The siliceous deposits show every gradation from a loosely
cemented gravel-conglomerate to a fine-grained and intensely

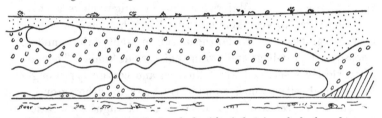

Fig. 31. Surface quartzite (marked with circles) in a bed of sand,
overlain by soil (dotted).

hard quartzite of a white, grey or yellowish colour; some
specimens are even bright green. Sometimes the sand-grains
are readily distinguishable; sometimes the cementing material
is deposited in crystalline continuity with the grains, so that
the outlines of the latter are completely invisible. The rock
often has a conchoidal or splintery fracture and is sometimes
almost like a chert in texture. The hard bed, which may be
as much as 10 feet thick, passes downwards by imperceptible
gradations into loose sand or clay. The hardened portions also
frequently form large flat masses lying isolated in a sandy or
argillaceous matrix or even in laterite. Surface quartzites are
very common in the south-western part of Cape Colony and also
in parts of Bechuanaland[1].

[1] Rogers and Du Toit, *Geology of Cape Colony*, 2nd edition, 1909, pp. 378–385.

Large areas in the central and northern portions of the Union of South Africa are covered by a calcareous deposit (surface limestone) whose origin is apparently somewhat similar to that of the quartzites just described. The rocks of the Karroo series contain a good deal of lime, and the calcium carbonate is formed by the weathering of these rocks. In many places the calcareous material attains a thickness of 12 feet[1]. It often forms a kind of pan immediately below the soil, and it is seen also in northern Cape Colony below a layer of red sand. In such cases it is often nodular in structure and may be compared with the kankar of India.

(b) *Cumulose deposits.* The second class of the sedentary deposits comprises all those accumulations that have been formed by the growth in place of vegetation, aided in some cases by the entangling of solid material among the stems and roots of the plants. Material of animal origin plays a small part compared to vegetable matter, although it may be locally of some significance. By far the most important of the cumulose deposits are peat-bogs and swamps of various kinds. They are often closely associated with alluvium and certain types might be assigned with almost equal correctness to either group.

The term *peat* is of somewhat wide and vague application, being generally applied to all those soils and surface deposits that are abnormally rich in humus and fibrous vegetable matter in a more or less advanced state of decomposition. The formation of peat is specially characteristic of cold temperate and arctic climates, and does not take place to any extent in the tropics. It is certainly brought about by a special type of bacterial action under definite conditions, the most important of these being saturation with water and a limited supply of air. Peat formation may therefore be described as limited oxidation, leading to the formation of compounds of the hydrocarbon group.

In Britain two types of peat are recognized, namely, *hill peat* and *fen peat*. Hill peat is chiefly found at considerable heights above sea-level in mountain and moorland regions; only in the

[1] Hatch and Corstorphine, *The Geology of South Africa*, 2nd edition, 1909, p. 329.

moist climate of western Scotland and of Ireland does it descend
to sea-level. Hill peat consists mainly of mosses, especially
Sphagnum moss, mixed with cotton-grass, rushes and roots of
heather. The upper layers are generally somewhat brown and
fibrous, becoming more compact and darker in colour downwards. The lowest layers are often quite black and
homogeneous, approaching lignite in character. Roots and
trunks of trees, often quite black, are commonly found buried
in peat bogs, indicating the former existence of forests in what
are now treeless regions. Of late years the peat bogs of Scotland and northern England have been made the subject of
careful investigation from the botanical point of view. It has
been shown that the growth is extraordinarily slow, the lower
layers of some of the Scotch peat-mosses dating back to the
later stages of the glacial period, and containing remains of
truly arctic plants. Interstratified with these are so-called
"forest beds" indicating a climate somewhat warmer than at
present, when trees grew at much higher elevations. Although
the conditions in Scotland now appear eminently favourable
to the formation of peat, it has been shown that its growth is
practically at a standstill, and, indeed, in places it is even being
destroyed by denudation[1].

Fen peat on the other hand is found at low levels in flat
marshy regions, such as the Fenland of eastern England, the
"moors" of Somerset (Sedgmoor and others) and certain parts
of central Ireland. It is a substance of a much more muddy
character than hill peat, consisting mainly of the remains of
water plants, such as rushes, sedges and grasses, with sometimes
abundance of moss (especially *Hypnum*). Interstratified with
the peat of the Fenland are layers rich in remains of forest trees.
Near Ely five such forest-beds have been found, one above the
other, each characterized by a different species of tree, and a
similar succession has been observed in the peat bogs of Denmark. The occurrence of forest beds in the fens can be explained
in two ways; either by the prevalence for a time of a somewhat
drier climate, or by a slight elevation of the land, leading to

[1] Lewis, "The Sequence of Plant Remains in the British Peat Mosses,"
Science Progress, vol. II. 1907, p. 307.

more effective natural drainage. Since the forest beds of Scotland are known to be due to climatic changes, those of the Fenland may probably be attributed to the same cause.

The greatest development of peat is to be found in the frozen regions of the Arctic zone, in northern Russia, Siberia, Alaska and northern Canada, constituting the Tundra. Here the annual average temperature is very low, the ground remaining permanently frozen below a depth of 3 or 4 feet, the limit to which the warmth of summer can penetrate. These conditions are specially favourable to the peculiar type of limited decomposition that results in peat. The dominant plants are mosses, especially *Sphagnum*, together with dwarf birch and willow, *Empetrum* and other flowering plants. The whole surface of the ground is composed of peat, usually in undulating, hillocky masses, with only occasional outcrops of bare rock. The Tundra may be considered as absolutely value-less agriculturally, being quite incapable of cultivation, or of maintaining any animals except the reindeer and the musk-ox[1].

Peat accumulations of various types cover very large areas in Germany, Holland, Denmark and the Baltic provinces of Russia. They are of very great commercial importance and are now exploited on a large scale as fuel, for moss-litter and for many other economic purposes. The literature of the subject is enormous and many varieties have been recognized, based mainly on differences in the botanical composition of the peat. In Germany, as in Britain, there appear to be two main types, which may for convenience be called dry peat (Trockentorf) and swamp peat. The term *dry* here is however only of relative value. The dry varieties include such forms as birch peat, pine peat, heather peat, bilberry peat and so forth, while the swamp varieties are named arundinetum, cyperacetum, hypnetum, according to the dominant species of grass, sedges and mosses. The common sphagnum peat, which is mainly used for the preparation of moss-litter, appears to be a somewhat intermediate form[2].

[1] Ramann, *Bodenkunde*, 3rd edition, Berlin, 1911, p. 578. Glinka, *Die Typen der Bodenbildung*, Berlin, 1914, pp. 236–242.

[2] Ramann, *Bodenkunde* 3rd edition, Berlin, 1911, pp. 171–186 and 231–238.

As before stated, the essential condition for peat-formation is slow decomposition of vegetable matter with a limited supply of oxygen at a comparatively low temperature. This can only be attained when the whole mass is saturated with stagnant water. Free circulation of water hinders the process, leading to free oxidation and probably to a different kind of bacterial action. Hence peat-formation always indicates deficient drainage and excess of ground water, which may be frozen, as in the Tundra. A heavy rainfall is of course a contributing factor of importance, but this alone will not give rise to growth of peat, as shown by its cessation in western Scotland, where the climatic conditions appear to be ideal. In all probability the temperature here is somewhat too high, and it is clear that a subarctic climate is the most favourable[1].

The chemical composition of peat is as a rule remarkably constant, as shown in the following table; the most variable constituent is the mineral matter, or *ash*; even when this is present in considerable quantity it is often deficient in potash and lime, and hence contains little available plant food. On the other hand iron is abundant, and sometimes gives rise to deposits of iron-ore in connexion with peat bogs. The formation of pans is also a well-known phenomenon (see p. 140).

	Thésy, France	Forest peat	Peat, Sweden
Carbon ...	50·67	51·47	51·38
Hydrogen ...	5·76	5·96	6·49
Oxygen ...	34·95⎱	32·68	⎱35·43
Nitrogen ...	1·92⎰		⎰1·68
Ash	6·70	9·67	5·02

On the whole lowland peat is much richer in ash than hill peat; according to Ramann[2] the amount in the former averages 10 per cent., in the latter 2 per cent. The ash in each case is constituted as follows:

	Nitrogen	Lime	Phosphoric acid	Potash
Hill peat	0·8 %	0·25 %	0·05 %	0·03 %
Fen peat	2·5	4·0	0·25	0·10

[1] Arber, *The Natural History of Coal*, Cambridge, 1911, pp. 54–64.
[2] *Bodenkunde*, Berlin, 1911, p. 234.

The differences here are very notable, and are quite sufficient to account for the comparative fertility of the peaty Fenland under cultivation. The hill peat is strangely deficient in all mineral plant foods. These figures emphasize strongly the need for potash manures on peaty soils in general.

With regard to true swamp deposits little information is available. Many swamps are the silted-up beds of lakes and rivers; hence they come more correctly under the heading of alluvium, since much of the material has been transported. This also applies to a large extent to the mangrove swamps of tropical coasts and still more forcibly to river deltas, where swamps are characteristic. The so-called swamp soils of the United States, allowing for differences of vegetation, appear from the published descriptions to be very similar to the lowland peat or fen peat of Europe. In the silting up of shallow lakes and pools with no outlet, or with merely small streams running through them, microscopic organisms, both vegetable and animal, seem to play a considerable part in the earlier stages, often forming a floating crust that gradually grows thicker and eventually forms a soil suitable for the growth of higher plants. In all such instances the final product is of a peaty nature, with a very high proportion of humus, and a small amount of inorganic material mainly derived from the ash or mineral constituents of the decayed plants.

Transported deposits. Under normal climatic conditions and more especially in temperate and cold regions, by far the greater part of the superficial deposits have undergone more or less transport; frequently the whole mass has been brought from a distance and may reach a great thickness. The geological agents here specially involved are running water, ice and wind. Each of these gives rise to a special type of deposit, needing separate consideration. Gravity is of minor importance and local in its action, depending solely on the configuration of the ground.

Transported deposits may be classified on the basis of the geological agent chiefly concerned in their formation, as follows[1]:

---

[1] Merrill, *Rocks, Rock-weathering and Soils*, 1897, p. 300.

Colluvial	...	Screes and cliff debris.
Alluvial	...	River and swamp deposits, loess (in part).
Glacial	...	Moraines, drumlins, boulder clay, gravels.
Aeolian	...	Wind-blown sand, loess (in part).

The only term requiring explanation is colluvial. This is derived from the Latin *colluvies*, a heap; all the other terms are in common use, and their significance is obvious.

Colluvial deposits. These are of little agricultural import- ance, being mainly found in uncultivated areas. In temperate regions they are mainly confined to mountainous situations, forming screes at the base of precipices and steep slopes generally.

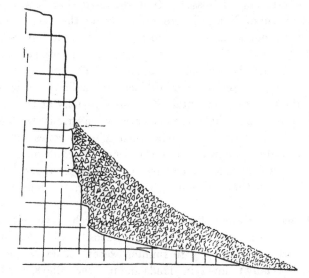

Fig. 32. Formation of a scree at the foot of a precipice, consisting of angular fragments of rock, shattered by frost-action or by changes of temperature. Such a deposit, when consolidated, forms a breccia.

In the case of screes the transport is mainly due to gravity, although in some instances running water also plays a part. The greatest development of screes is undoubtedly in arid regions, such as Baluchistan and the Sinai peninsula. Here the main agent of disintegration is change of temperature. In colder climates frost acts in a similar way (e.g. the well-known screes of the Lake District).

In certain parts of the south of England, beyond the southern limit of the Pleistocene ice-sheet, considerable areas are covered by accumulations which may conveniently be placed in this category. Such are the *Coombe Rock* or Elephant-bed of the South Downs and the *Head* of Cornwall. The Coombe Rock is found in the dry valleys or coombes of the Downs, and consists of a thick mass of angular flints, with more or less loamy matrix. It contains in abundance broken and decayed teeth of horse and elephant. When traced to the lower ground the larger elements become relatively less abundant, and the Coombe Rock appears to pass laterally into the brick-earths of the coastal plain of Sussex. It is supposed that this deposit and the Head of Cornwall were formed during the glacial period, when the ground was frozen and the rainfall could not sink in, as it now does, but flowed over the surface in torrential streams[1].

Many water-borne deposits might also be assigned to this class, as for example the piles of blocks and debris so often seen along the courses of mountain streams, especially at the points where they reach the lowlands or where a tributary joins a main valley. These are often called alluvial cones, but the name is scarcely appropriate in this sense. Such deposits may, in course of time, become covered by finer material and by soil and they sometimes give rise to good pasture in mountainous regions.

Alluvial deposits. This is a very comprehensive term and includes a great variety of superficial deposits, all of which are due directly or indirectly to running water. They cover vast areas, especially in low-lying lands and in river valleys, and they constitute the subsoil of some of the most fertile regions of the world. As examples of alluvium on a large scale special mention may be made of the Mississippi valley, the deltas of the Ganges and Bramaputra, of the Nile and other African rivers, and the great plains of China. Delta deposits may perhaps be regarded as the typical alluvium, but the flood plains of large rivers are also very important, while along the courses of nearly all rivers, at any rate in their lower reaches, more or less alluvium is

---

[1] Reid, *Quar:. Journ. Geol. Soc.* vol. XLIII. 1887, p. 364. Elsden, "Geology in the Field." *Geol. Ass. Jubilee Volume*, 1910, p. 275.

found. Fertile alluvial plains are also formed by the silting up of lakes.

Alluvial deposits vary enormously in character, ranging from piles of immense boulders along the courses of mountain streams, through all grades of size down to the finest silt and warp of tidal rivers. They may therefore be referred to all the normal categories of sediment; boulders, gravel, sand and mud. A distinctive feature of most alluvium is a high proportion of organic matter, and many alluvial deposits might almost equally well be classified as peat and swamp soils.

The coarse-textured boulder deposits of torrential streams in hilly regions are of little or no agricultural importance and may be disregarded. River gravels of moderately coarse grain, with abundant pebbles of 2 or 3 inches in diameter, cover very considerable areas in comparatively low ground, and are of some importance, as for example along the chief rivers of southern England, and especially the Thames. The gravels of the Thames are clearly divided into terraces, at different heights above the present river level, and the same statement applies to the gravels of the Great Ouse, the Cam and other rivers of the eastern counties. The generally accepted explanation of these terraces is that they belong to periods when the land stood at different heights above sea-level. The oldest and highest, the 100 foot terrace of the Thames, was formed by that river when the land was generally about 100 feet lower than at present. When an uplift occurred, the equilibrium of the river was disturbed and it was forced to deepen its channel, in the endeavour to establish the base-line of erosion proper to the new conditions. The new valley was narrower than the old one and some of the alluvium of the older period was left as a terrace on the sides of the valley. A repetition of this process gave rise to a still narrower valley and formed the 50 foot terrace. From a study of the flint implements and other human relics found in the gravels it has been found possible to correlate the formation of such gravel terraces with different stages of culture (Palaeolithic man), and with the occurrence of certain types of extinct animals, especially different species of elephant, rhinoceros and hippopotamus. Fig. 33 shows the relation of such gravel terraces to each other and to the under-lying older strata[1].

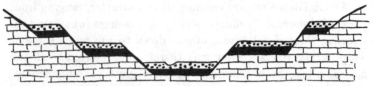

Fig. 33.   Gravel terraces along the course of a river.   The figures show two
terraces and the gravels formed at the present river level.

Besides the gravel terraces clearly belonging to earlier stages of existing rivers, there are also in the eastern counties obvious river-gravels formed by rivers which have long ago disappeared. Some of the best known and most interesting of these are in Cambridgeshire. The gravels of the ancient river system start in dry valleys in the Chalk near Newmarket; they can be traced as low ridges over the lower ground, crossing the present Cam at right angles just below Cambridge and running towards St Ives. Near March they can be followed laterally into marine gravels with shells of a somewhat arctic character, showing that at the time of their formation the sea extended far into what is now the Fenland. All these gravels consist mainly of flints, with a very small admixture of other stones; indeed on all Chalk tracts the conditions are specially favourable to the formation of river gravels, owing to the abundance and durability of flints. The pebbles of the Thames gravels are partly derived directly from the Chalk, but also largely in an indirect manner from the Tertiary strata. In Norfolk and Cambridgeshire the gravels contain far-travelled stones derived from the glacial drift and originally transported by the ice from northern England, Scotland and Norway. River gravels also cover large areas in the Trent valley, especially near Newark, extending as far east as Lincoln. It is believed that the Trent once flowed through the gap in the Jurassic escarpment at Lincoln, and that its present course by Gains-borough to the Humber is of comparatively recent date. The presence of these gravel deposits helps to confirm this view.

[1] Salter, "On the Superficial Deposits of Central and Parts of Southern England," *Proc. Geol. Ass.* vol. XIX. 1905, p. 1.

The majority of river deposits in low-lying and cultivated regions are of finer grain than the foregoing, ranging from sand through loam to silt, and in estuaries even to mud. These constitute the basis of some of the most fertile of all known soils, especially along the lower courses of the great rivers of tropical and subtropical countries. In temperate and cold climates river alluvium often shows a peaty tendency, but in the tropics oxidation and bacteria are too active for the growth of peat to attain its full development. The nature of the alluvium brought down by any given river must naturally depend on the character of the rocks over which it flows, hence infinite variety is possible, and since few rivers flow solely over one kind of rock, alluvium is generally of a very mixed nature. This accounts to a large extent for its general fertility, since all the constituents of plant food are likely to be present.

The best examples of alluvium are to be found in countries that have not been glaciated in geologically recent times. Within the area overrun by the Pleistocene ice-sheet there is always much glacial material mixed with the alluvium, and this is specially the case in the British Isles. North of the Thames it is almost impossible to find alluvial deposits free from far-travelled ice-borne material, and even in the southern counties it is yet doubtful to what extent ice played a part in the formation of the superficial deposits. The great "alluvial" tracts of the Fenland and of the Humber estuary are not true river alluvium, but consist mainly of marine silt and peat (see pp. 119 and 128).

The delta of the Nile affords a particularly good example of alluvium, attaining a great thickness, as revealed by deep borings. The deposits consist of irregular alternations of sand and mud, varying in texture according to the conditions of deposition. The larger sand grains are much rounded, showing derivation from the desert, but as usual the smaller grains are more angular. The mud consists essentially of the same materials as the sand in a finer state of division, and there is in addition a varying amount of organic matter. Owing to the annual floods of the Nile, the conditions are here specially favourable to the accumulation of alluvium over a very wide

area and the soil produced from it is also remarkably
fertile.

Along the lower course of the Mississippi there is a great
spread of river alluvium which yields a good soil. The
Mississippi and Missouri are both remarkably muddy rivers,
and it has been estimated that they carry down to the sea
annually more than 7,000,000,000 cubic feet of solid matter in
suspension. Consequently the delta is growing very fast and
now extends a long way into the Gulf of Mexico. The flood
plain of the Mississippi is several hundred miles in length and
many miles wide; on either side it is bounded by steep bluffs
and the area between these is constantly flooded by the river;
thus receiving a fresh coating of finely divided sediment
which is added to the soil and increases its fertility. Much of
the flood plain is covered by thick forests and undergrowth,
but there are also considerable grassy areas. Sometimes,
owing to accumulation of sediments, the streams naturally
flow at a higher level than the general surface of the ground
this feature being often accentuated by artificial banks or
"levees." In violent floods the rivers sometimes change their
courses, leaving long, narrow, partly dried-up channels; these
are called "ox-bow lakes." The great alluvial plains of the
Ganges and of the rivers of China are very similar in their
general characters and are, for the most part, extraordinarily
fertile.

The alluvium of the English Fenland is of a rather special
character. Although several fairly large rivers run into the
Wash, the Fenland is not in the main a delta deposit in the
ordinary sense, since the material has been to a large extent
brought in from the sea. The growth of peat also plays a
considerable part (see p. 119). The deposits of the Fenland are
of three different types, namely, (1) gravel, (2) marine silt,
(3) peat. The gravels come to the surface furthest inland,
forming a continuation of the river gravels and probably
underlying the other deposits also; next comes the peat, while
the silt is now chiefly seen at the seaward margin. However
the peat of the landward region is nearly everywhere underlain
by silt and the different types often alternate vertically as well

as horizontally. The marine silt occurs in two different types; one is a clay, the so-called *buttery clay*, while the more sandy type is often called warp (compare the warp of the Humber, to be described later)[1]. The prevalence of marine silt can be explained as follows. The coasts of north Lincolnshire and south Yorkshire consist of soft strata, mainly glacial, which are being rapidly eroded by the sea. The set of the tidal current is here from north to south, running strongly into the opening of the Wash, which formerly extended at least as far as March and perhaps further. Hence the fine silt carried by the tide was deposited in the slack water, and gradually filled up the bay. When it had reached a certain height the growth of peat began and completed the process. The sandy warp forms a fertile soil which of late years has been found admirably suited to bulb-growing. This is now quite an important local industry around Spalding. Another crop peculiar to the Fenland is woad, which is still grown to a small extent near Wisbech.

The waters of some rivers are specially rich in suspended material and these are most liable to form alluvial deposits. The Humber is said to be the richest in silt of all the British rivers, and advantage is taken of this fact to increase the fertility of the adjoining land by artificial means; the process is known as *warping*, and may be described briefly as follows. Warping consists of letting in silt-laden tidal water through sluices and excavated drains, and allowing it to stand upon the selected lands, the water being gradually drawn off again with the fall of the tide; the silt or warp is thus deposited, and the process, when continued for some time, results in the formation of a soil of great fertility. Some of the warp-lands of the Humber are said to be among the richest soils of England[2].

Glacial deposits. These constitute a large group of surface accumulations of very great importance in the British Isles, in North America and indeed in all those countries that came within the influence of the Pleistocene glaciation. Glacial deposits of recent age are also found in the higher mountain

[1] Skertchley. "The Geology of the Fenland," *Mem. Geol. Survey*, 1877. Rastall. "The Geology of Cambridgeshire, Bedfordshire and West Norfolk," *Geol. Ass. Jubilee Volume*, 1910, p. 177.

[2] *Third and Final Report of the Royal Commission on Coast Erosion*, 1911, p. 127.

ranges of almost all parts of the world, but these are of less importance than those of the earlier Pleistocene period. Glacial deposits display very great variety of composition, but they can in general be divided into two main groups, namely:

(1)   Sands and gravels,
(2)   Boulder-clay.

It does not come within the province of this chapter to deal with all the characteristics of glaciated regions; scratched surfaces, *roches moutonnées* and other features of glacial topography may be disregarded, as we are here dealing solely with deposits that may give rise to soil.

Glacial deposits may be laid down either by the ice itself, as moraines and other related structures, or by the water flowing from the ice. The latter originates many of the masses of sand and gravel, while boulder-clay is apparently always formed directly by the ice. The deposits of the streams of water flowing from the ice are now commonly designated *fluvioglacial*, and in recent times more and more importance has come to be attributed to this mode of origin.

In this connexion it must not be forgotten that for many years a controversy has raged as to the nature of the conditions that gave rise to the glacial deposits of Pleistocene age in Britain and elsewhere. On this question geologists are ranged into two schools, the one maintaining a general submergence of the whole area to a depth of many hundreds of feet beneath the waters of the sea; this school considers that the main work of transport and deposition was performed by floating ice, as bergs and floes. The other school maintains that the relative levels of land and sea were then much the same as now, and that the whole area was covered by an ice-sheet like that of Greenland, the motive power of the ice being its own weight; the vast accumulation of snow and ice at the centre forced the outer parts to move for great distances, even across the sea and over hills of considerable altitude. Thus on this theory Scandinavian ice must have crossed the North Sea and climbed to a height of several hundred feet on the coast of Yorkshire. In spite of the great mechanical difficulties involved, the ice-sheet theory

is now accepted by the majority of geologists. One thing at
any rate is universally admitted, namely, that during the
glacial period there were glaciers of the alpine type in the
mountains of Britain, and that these formed moraines and
other structures such as can still be seen in regions where
glaciers now exist. The main controversy is concerned with the
origin of the glacial deposits of the midland plain and of the
eastern counties.

The essential feature for the present purpose is that vast
areas of land in Britain and elsewhere are covered by a thick
layer of *drift*, undoubtedly of glacial origin and containing
material that has been transported for great distances. This
drift varies much in lithological character and in thickness,
but it is usually sufficiently thick to exert a preponderating
influence in determining the character of the soil.

The exact nature of the glacial deposits of any region is
naturally determined by the source of the material composing
it. In most instances a considerable part of the drift is derived
from the underlying rocks, but there is always more or less
admixture of material of distant origin. Thus the glacial
gravels of Norfolk and Suffolk consist mainly of flints, while
most of the boulder-clay is chalky. In Cheshire and Staffordshire the drifts are largely made up of Triassic material, with
pebbles from the Bunter pebble beds and abundant rock-fragments from Scotland and the Lake District. The boulder-clays
of the Midlands are made up of debris from the different Jurassic
strata, with usually a good deal of Chalk from Lincolnshire.
Many other instances might also be given, but it will perhaps
be more satisfactory to describe the sequence of the drifts in
one or two typical areas as examples.

From a study of the drifts of Norfolk, Suffolk and Cambridgeshire it has been found possible to establish a definite
sequence, as follows:

     5. Plateau gravels.
     4. Great Chalky Boulder-clay.
     3. Mid-glacial sands.
     2. Contorted drift.
     1. Cromer till.

Of these five subdivisions only the third and fourth are of much importance agriculturally. The Cromer till and the contorted drift are hardly seen except in the cliffs of the coast. The latter however is supposed to form the subsoil of a very fertile area near Norwich. The Cromer till is a heavy brown clay and the contorted drift is a light yellow loam, stratified and extraordinarily contorted, with blocks of Chalk up to one or two hundred yards in length. Both divisions contain many Scotch and Scandinavian rocks and were clearly formed by ice coming from the North Sea. Glacial sands and gravels cover large areas of the three counties named; they vary from coarse flint gravels to fine sand, the latter often being much wind-blown. Owing to the prevalence of these deposits there is much poor sandy land in the west of Norfolk and of Suffolk, a large area being uncultivated and under heather and bracken. This kind of land is of more sporting than agricultural value. The glacial gravels are often difficult to distinguish from the river gravels before described. Some of them probably belong to the mid-glacial series, but part are undoubtedly newer than the Chalky Boulder-clay.

The Great Chalky Boulder-clay is the most widespread and the most important of all the drift deposits of eastern and central England. Although varying a good deal in character it usually contains more or less Chalk, together with far-travelled boulders, which are sometimes derived from older glacial deposits. The nature of the boulder-clay is largely determined by the kind of rock on which it lies, since much of the material has not been transported very far. The argillaceous portion is mainly derived from the clays of the Jurassic and Cretaceous, and the colour and texture of the boulder-clay are determined by this fact; consequently also derived Jurassic and Cretaceous fossils are common. From a study of the distribution of the different types of boulder-clays and their contents in eastern and central England it has been found possible to obtain much information as to the sequence and direction of the ice-streams of the Pleistocene glaciation. It is clear that by far the greater part of the material constituting the drifts of eastern England is of local origin, the foreign portion, though the most noticeable, being really insignificant in total amount.

The plateau gravels of eastern England have been previously mentioned and their manner of formation described (see p. 113). Such gravels of glacial or fluvioglacial origin cover large areas in Norfolk, Suffolk and Cambridgeshire and give rise to land which is in places almost completely uncultivated, being mainly under bracken and heather, and sometimes entirely barren. Similar spreads of high level gravels form many of the open commons in the home counties north of the Thames. Some of these are now becoming of value as residential districts, especially in Buckinghamshire.

The glacial deposits of Lincolnshire and Yorkshire resemble those of East Anglia in general character, though differing in detail. The drifts of Holderness have been exhaustively studied by the Geological Survey[1]. It is believed that some of the clay has here been redeposited by wind (see also pp. 136 and 139).

The drifts of western England differ a good deal from those of the east, since they are formed from very different material. On the plains of Lancashire and Cheshire are found boulder-clays rich in material derived from the Trias, especially from the red marls of the Keuper. Rounded sand grains and pebbles from the Bunter are also recognizable. The far-travelled boulders come from various sources; from the Lake District and Galloway and also from the mountains of North Wales. The latter are abundant as far east as Birmingham. Pebbles from the Bunter pebble bed are found scattered over the surface in many parts of the Midland counties, showing the former existence of a thin covering of drift which has now mainly disappeared or become incorporated with the soil.

In almost all the hilly and mountainous regions of the British Isles glacial deposits are very largely developed as moraines, drumlins, eskers, gravels and boulder-clay. In fact it may be said that in such regions they constitute the commonest subsoil. They vary greatly in lithological character and yield soils of widely differing character according to the

[1] "Geology of Holderness," *Mem. Geol. Survey*, 1885.

kind of rock forming the gathering ground of the glaciers. The soils are usually stony and not very fertile.

The thickness of glacial deposits varies enormously, ranging from a mere surface skin to several hundred feet. It is evident that a very small thickness of drift is sufficient to determine the character of the soil. Hence the importance of the study of drift maps. Sometimes it is shown by well-borings that pre-glacial river valleys have been filled by drift to surprising depths; at Glemsford, near Sudbury in Suffolk, a boring passed through 470 feet of drift, the floor of the old valley being nearly 350 feet below present sea-level. Near Newport, Essex, 340 feet of drift was passed through and in the north of England records of 200 feet of drift are not uncommon. All this indicates that in pre-glacial times the land stood much higher than now, and the area now occupied by the North Sea was then in all probability an alluvial plain formed by the Rhine and its tributaries. A knowledge of the existence of such deep drift-filled channels is evidently of great importance in questions of water-supply, but unfortunately they can only be detected by actual boring.

Deposits formed by wind. Deposits of this class are naturally most abundant in dry climates, and in temperate regions with considerable rainfall they are of but local and limited occurrence.

In our own country the most conspicuous wind deposits are the sand-hills of the coast. Owing to the prevalence of westerly and south-westerly winds they are much more prevalent on the west coast of Britain than on the east; still they are developed to a considerable extent on certain parts of the eastern coast, especially in Norfolk and in many parts of Scotland, e.g. Fife, Aberdeen, and the shores of the Moray Firth. On the west coast, wherever the shore is low and not bounded by cliffs, sand-hills are almost universal. The width of the belt of dunes is not generally great, not often exceeding a few hundred yards. The primary source of the material is in nearly all instances the sand of the beach; this is cast up by the waves and carried inland in large quantities by the prevailing winds. Coastal sand-hills are of little or no agricultural value, but are now very commonly utilized as golf-links. The sand-hills of the coast are not always stationary, but often tend to encroach on cultivated ground, sometimes to a disastrous extent. This state of things may be seen in many places in England, especially in Cornwall and Suffolk, but one of the most notable examples was the destruction of the Barony of Culbin, on the southern shores of the Moray Firth. Several thousand acres of the most fertile land in the county of Elgin were suddenly overwhelmed by sand in the seventeenth century and have remained in that condition ever since. In many cases such movements of sand-hills have been promoted by destruction of trees growing on them, and conversely it has been found that such destructive movements can be prevented by afforestation, especially by the planting of pine-trees, as has been largely done in Norfolk, on the Holkham estate and elsewhere.

The most extensive development of wind-blown sands is found in the true desert areas of the arid zone; here however agriculture in the true sense is non-existent. There is no doubt that within historic times the advance of desert sands has overwhelmed fertile and highly cultivated areas, as in parts of the Nile valley and over a large region in south-western Asia.

It is however in South Africa and Australia that wind-borne deposits attain their greatest practical importance. Where the climate is hot and dry changes of temperature and wind are the chief geological agents, and the main product of their activity is sand. In such areas the vegetation is specially characterized by the abundance of succulent plants, capable of resisting prolonged drought and the agriculture therefore shows special features, which cannot here be described in detail. Normal arable farming can as a rule only be carried on where alluvial deposits occur, or where artificial irrigation is possible. There the soil is often remarkably fertile, but the unwatered areas are only suited to pasture.

As before mentioned there are certain limited areas in the British Isles where wind transport plays a part in the formation of superficial deposits. Where glacial sands are abundant, as in parts of Norfolk and Suffolk, they are often much blown by wind and it is believed that the loamy soils of Holderness were partly formed by wind-transport of the finer argillaceous portions of boulder-clay.

One of the best known of all wind-borne formations in temperate regions is the Loess, which covers such large areas in central Europe and in Asia, from the shores of the English Channel to China. It is probable that not all the deposits included under the name of Loess were formed in the same way, but the greater part is generally believed to be due to wind. During the interglacial periods and after the final departure of the ice, the glaciated area and the regions immediately bordering it must have been largely covered by glacial mud of very fine texture, laid down by streams flowing from the ice. This mud was formed from the fine "rock flour" ground from the surface of the rocks by the passage of the ice, such as may be seen to-day imparting a milky tinge to all rivers flowing from glaciers. When the climate became warm and dry the mud was converted into dust and transported far and wide by the wind, as can now be seen in central Asia, where mountain chains are partly or almost completely buried in wind-borne dust. The Loess is normally a fine yellowish loam of very uniform texture, often traversed by capillary tubes lined with calcium carbonate and due to roots. These partly account for the facility with which it forms vertical cliffs when subjected to stream-erosion. The Loess often forms a soil of remarkable fertility, and when mixed with a certain proportion of peaty matter it gives rise to the well-known Tchernozom or Black Earth of Russia (see also p. 148).

Loess is also extensively developed in certain parts of the United States, especially in Minnesota and Dakota, and also in Manitoba. The so-called Adobe of the Mississippi valley is very similar and has a like origin.

CHAPTER VI

SOILS

In the preceding chapters we have considered in a general way those parts of geological science which are specially applicable to agriculture. It now remains to combine the information there contained into a more specialized account of the geology of the soil, using this word in the somewhat narrow sense as generally adopted by agricultural writers and by practical men. In this sense the soil is taken to mean in arable land the cultivated layer down to the depth to which ordinary implements penetrate, and in pastures down to the usual limit of root growth. The more or less weathered and disintegrated portion below this is called the subsoil; this part is not disturbed during ordinary agricultural operations, but crops obtain a good deal of their food-supply from it. The subsoil passes downwards by indefinite gradations into the rock below, and in different places it varies very greatly in thickness, being sometimes almost completely absent.

It has already been made apparent that soil-formation is a very complicated process, involving a large number of different agencies. These may as a matter of convenience be divided into three groups, chemical, physical and biological, but all of these are mutually interdependent, and as a rule it is impossible to disentangle the effect of each.

The first step in soil-formation is weathering; by this term we understand disintegration of rocks, usually accompanied by chemical and mineralogical changes in their component minerals. It is however possible for a rock to be broken up by purely physical means, without any chemical change.

This occurs chiefly in arid or very cold regions, where water
has no effect. In moist climates water is always present and
brings about important changes, both directly by solution and
indirectly by favouring oxidation and chemical reactions in
general. The other principal agents of weathering have
already been enumerated in Chapter II. The general effect of
all these taken together is to produce on the surface of the ground
a layer of loose and disintegrated material, which in itself
constitutes soil. It is evident that in such a case the nature
and composition of the soil is controlled to a large extent by
the original composition of the rocks from which it is derived
and the nature of the chemical processes that have taken place
during its formation. Chemical processes may remove some
of the elements originally present, but in general they cannot
add to the number, except when aided by transport of material
in solution, or by some similar method. Soils freshly formed
in this way by simple weathering of rocks are therefore often
deficient in some of the constituents of plant food, while those
that are present may not exist at first in an available form.
In particular the supply of organic matter is at first small and
will gradually increase in course of time as animals and plants
die and decay in the soil. Hence all soils both natural and
cultivated tend to get richer in organic matter or humus, and
consequently in nitrogen.

In the majority of instances the weathered material does not
remain in exactly its original position. Nearly always more or
less transport takes place, as before described, and the constituents of many soils have been derived entirely from distant
sources. Consequently they are often of a very mixed nature,
and have no necessary relation to the underlying rock. These
facts afford a basis for a classification into *sedentary* and *transported* soils. As examples of sedentary soils of widely differing
type we may refer to the Chalk soils of the Downs of the south
of England and the peaty soils of the Fenland. Good examples
of the transported class are afforded by the drift deposits of
glacial origin and the alluvium of many lowland rivers. In
many examples of the latter group the difference between soil
and subsoil is very small.

Modification of soils by wind-transport. Some 30 years ago attention was called by Mr Clement Reid[1] to some anomalies in the character of certain soils when compared with the substratum on which they lie. It has long been recognized that in arid countries transport of dust by wind can give rise to important superficial deposits, as in the case of the Loess, and Darwin[2] also called attention to this possibility. There are certain areas in the east of England where the rainfall is small, especially in spring, and at this season strong winds are prevalent, especially from the east; these have a powerful desiccating effect on the superficial deposits. Mr Reid quotes an instance of a field near Cromer which was sown three times in one spring and finally left fallow, as the whole of the soil was banked up like a snowdrift against the hedge. Similar effects can be seen frequently in the cultivated parts of the Breckland of western Norfolk, where it is worse than useless to cultivate the land on a windy day. In course of time this process of sifting by wind must bring about considerable changes in the character of the soil by removing the finer particles, the clay constituents, leaving the heavier sand behind and thus making the soil still lighter (see also p. 133). During the examination of the superficial deposits of Holderness and Norfolk by the Geological Survey it was found that the more high-lying areas of boulder-clay carried a comparatively light loamy soil, while large areas of almost pure sand and gravel were also covered by fertile loams. On the other hand in the hollows the soils were extraordinarily fine in texture, even where the substratum was not specially heavy. This is probably due to removal of fine particles of clay by the wind from exposed places with deposition in sheltered spots. Much of the sandy soil covering the Chalk Wolds of Yorkshire and Lincolnshire may also be due to wind-transport.

It is improbable that wind-transport has much effect on the soils in the western and northern portions of our islands, or in fact in any region where the rainfall exceeds 30 inches or

---

[1] Reid, "Dust and Soils," *Geol. Mag.* 1884, p. 165, also "Geology of Holderness," *Mem. Geol. Surv.* 1885, p. 115.

[2] Darwin, *The Formation of Vegetable Mould*, p. 236.

thereabouts, but in the dry regions of eastern England there can be no doubt that it has an important influence on the texture and fertility of the soil, this effect on the whole being one of amelioration, making light soils heavier and heavy soils lighter, thus giving rise to a general levelling-up. It is only injurious in extreme cases, especially where the soil is removed bodily.

Formation of " pans " in soils. Almost all writers on agriculture have described the formation in soils of hard layers, or "pans," a short distance below the surface, but few of them appear to have realized that two different phenomena are here confused. The simplest kind of pan is of purely mechanical origin, and is confined to soils that have been long under cultivation. The manner of formation is as follows: year after year the plough penetrates to the same depth, turning over the upper layer, from 7 to 9 inches thick, according to the character of the soil and the ease of cultivation, but tending to consolidate the layer below, which has to bear the weight of the plough, and must become somewhat more solid every time it is compressed by the weight of a horse or of an implement. Ultimately this is pressed and trampled hard, and it is to be noted that in this country the loosening effect of winter frost does not penetrate more than a few inches into the soil. This hardened and compressed layer forms a barrier to the free passage of air, water and the roots of plants. A pan of this sort is formed most readily in clay soils, and is unknown in light sands. The obvious remedy is occasionally to cultivate the soil to an extra depth, or to employ the subsoil plough; these are unfortunately somewhat costly remedies, but will often repay the extra expense and labour involved.

The other kind of pan, the Ortstein of the Germans, is of somewhat more complicated origin, and is due primarily to the circulation of soluble material in the soil. Such pans are most commonly formed in soils rich in humus, in regions of abundant seasonal rainfall, especially where the soil is of a peaty nature. In the upper layers of such soils the vegetable matter undergoes a special kind of decomposition, giving rise to humous compounds which generally have an acid reaction, the so-called humic acids. By the weathering of minerals also many soluble substances are formed, chiefly salts of iron, lime, magnesia and other bases. These are carried down in solution into the subsoil and underlying rocks, so long as the general movement of water is downwards. Where there is free drainage at deep levels, that is, where the underlying rocks are porous, all these soluble matters are completely removed and there is no pan-formation. The chief result is a bleaching of the lower layers of the soil and of the subsoil. Such bleached layers are often to be seen below thin peaty soils resting on a sandy foundation. But in many cases the water is not carried away by free drainage, being held up by an impermeable layer, and the soluble materials tend to accumulate in the soil water. During the drier season of the year the downward movement of rain-water ceases; on the other hand the ground-water from the lower layers begins to move upwards to replace that lost by evaporation from the surface; hence the soluble materials also have an upward tendency. When the top layers of the soil are dry, atmospheric oxygen is abundant in the soil and the solutions as they rise are oxidized; this oxidation leads to a precipitation of various compounds, especially iron oxides and hydroxides and humus, in the soil, forming a solid layer, usually at a depth of 15 to 20 inches, this being the depth commonly reached by atmospheric oxygen.

The essence of the whole process lies in the fact that when soils rich in humus are saturated with water, *reduction* is taking place, whereas when the soils are free from interstitial water, and therefore full of air, *oxidation* is dominant, destroying the humic acids and liberating the iron, lime and magnesia. Under such conditions the liberated substances, oxides and hydrated oxides of iron, calcium carbonate, etc., may form either nodules and concretions in the soil, or in some cases a continuous layer or pan. The composition of these solid segregations varies widely; in some soils they consist of pure or nearly pure calcium carbonate, such as the kankar of the Indian cotton soils, and the surface limestones of South Africa. More commonly however they are composed of varying mixtures of iron hydrates with organic material (humus), the relative proportions varying with the humidity of the climate. In damp regions the pan may contain as much as 17 per cent. of organic matter; in arid regions the amount generally varies between 1 and 3 per cent.[1]

The Ortstein of German authors, which is so common and characteristic in the sandy soils of north Germany and Russia, may be described as a bed of sandstone cemented by humus rendered insoluble and hard by oxidation. The published analyses of this substance do not show any excessive amount of iron, but alumina is fairly abundant. When left exposed to the air the Ortstein quickly decomposes to a loose brown sand, owing to decomposition of the cementing material, and this breaking up is much facilitated by frost.

The "pans" formed in the soils of this country appear to consist in the main of hydrated oxide of iron (limonite). This is found both in sands and clays which are imperfectly drained and therefore waterlogged. The depth at which it occurs is doubtless controlled partly by the lower limit of cultivation, as previously explained, and partly by the distance to which air can penetrate freely to cause oxidation. Pan-formation in this country is a subject requiring much more complete and careful investigation than it has yet received, and in particular careful analysis of a considerable number of samples from different localities is most desirable. It has been suggested with a considerable degree of probability that bacteria, especially the iron-bacteria of Vinogradsky, may play a part in the process, but on this point no certain information is available[2].

In certain arid regions where the soil is rich in sodium carbonate (black alkali soils) a hard pan is often found, sometimes so hard as to be with difficulty broken with a hammer. The existence of this pan is due to the fact that sodium carbonate prevents flocculation of finely divided clay substance, but on the other hand converts it into a hard horny mass, quite impervious to water. This effect can be counteracted by neutralization of the alkali by treatment of the soil with gypsum[3].

---

[1] Treitz, 'Was ist Verwitterung?" *Comptes Rendus* 1ère *Conférence Internat. Agrogéol.* Budapest, 1909, p. 138.   Ramann, *Bodenkunde,* Berlin, 1911, p. 204.

[2] Hall, *The Soil,* 2nd edition, London, 1910, p. 285.

[3] Hilgard, *Soils,* New York, 1906, p. 62

Since the information on the subject of pans contained in
English agricultural literature is very scanty, here are appended
two analyses of such material carried out by the author. It
is seen that by far the most abundant constituents, besides
sand, are organic matter and iron oxide, although the latter is
in considerably less proportion than is generally believed to be
the case. In the first instance, a pan from Norfolk with a
highly ferruginous appearance, the iron oxide only amounts to
a little over 4 per cent.; nevertheless this pan is very hard and
presents a serious barrier to successful cultivation. The pan
from Cheshire, locally called "fox-bench," is very dark in colour,
almost black, and this feature is correlated with the very high
proportion of humus. The analyses were carried out on air-
dried material; other constituents though possibly present
were in amounts too small to be determinable.

	Brown ferru-ginous pan. Little Snoring, Norfolk	Black pan (Fox-bench) Delamere Forest, Cheshire
Organic matter	6·50	13·04
Sand and silicates (insoluble)	87·13	83·55
Iron oxide	4·31	2·82
Lime	·90	trace
Magnesia	—	·20
Phosphoric acid	trace	—
	98·84	99·61

The characters of soils. The agricultural value of a soil
depends on a large number of factors, only some of which are
within the province of the geologist. Pre-eminent among these
are its mineralogical composition and the state of division of
its particles; on the former subject little work has been done,
and a large field for research is here open, especially with regard
to the mineral character of the smaller particles. On the other
hand mechanical analysis by various methods has been made
the basis of soil-investigations by workers in many places, in
this country for example at Rothamsted and at Cambridge.
Owing to want of space the methods employed cannot be
described here. The study of the biology of the soil is a rather
large and controversial subject, on which also much remains

to be done, especially in the direction of bacterial investigation. The epoch-making discovery by Hellriegel and Willfarth in 1887 of the symbiotic nitrogen-fixing bacteria in the root nodules of the Leguminosae opened up a wide field of enquiry, of immense practical value, and furnished a scientific explanation of the long-observed fact that the leguminous crops enrich the land rather than impoverish it.

In this book no attempt is made to deal with certain properties of the soil of an essentially physical nature, such as texture, porosity, absorptive power for water and so on. These subjects, though of much practical importance, are not strictly geological, and are treated fully in many agricultural works[1].

Classification of soils. The ordinary practical classification is a somewhat rough and ready one, founded on the agricultural character of the soil, its more obvious characteristics and the comparative ease or difficulty of working, rather than on any very precise examination of its constituents. The practical farmer designates his soils as heavy or light, according to the difficulty or ease of working, and irrespective of actual specific gravity; in point of fact a cubic foot of sand is heavier than a cubic foot of clay. Here tenacity is really the determining factor, not weight.

A useful working classification of this kind is the following, which divides soils into six classes, namely, gravelly, sandy, loamy, marly, clay and peaty soils. Most of these terms explain themselves.

A gravelly soil is one containing abundance of stones, imparting a distinct character to it, while the term sandy certainly needs no explanation here. A loam is a soil containing sand and clay in approximately equal proportions, and for most purposes this is the best sort of soil, being rich in plant food and also free working. A marl is a soil containing a fair proportion of calcium carbonate in addition to clay; marls are often rather heavy, pasty soils. Peaty soils are those in which organic matter is present to an excessive degree; they are

---

[1] See especially Hall, *The Soil*, 2nd edition, London, 1910, Chapters III–v. This book also contains an excellent chapter on mechanical analysis.

generally dark in colour and often easy to work, though soft in wet weather.

Besides this rough and ready every-day classification, numerous attempts have been made to draw up schemes based on the geological origin of soils, on the climatic conditions controlling their formation, and on many other factors. Some of the most elaborate of these schemes are due to Russian workers, such as Sibirtzev and Glinka. Many others also exist in German and American agricultural literature.

In the most satisfactory of these classifications the basis is climatic, being really founded, though this is not always explicitly stated, on the primary subdivision of the globe into climatic zones. Here as in other cases it is necessary to bear in mind that an increase of altitude may produce the same effect as an increase of latitude, i.e. highlands in the tropics may have a temperate climate and the highest mountain ranges all over the world are more or less arctic. One of the most important of all the controlling factors is the amount of the rainfall, and the soils of arid regions show very special characters, which are not to a great extent dependent on latitude or altitude, not so much so at any rate as in moist climates.

Although climate is affected thus by many local variations of elevation, aspect and so forth, if a sufficiently large area is taken it is possible to draw up generalizations, and to divide a country into regions characterized by soils of a more or less definite character. The British Isles are too small in area, and built up of too many varieties of rocks, to afford a satisfactory basis, and the matter is rendered still more difficult by the varying relief of the land and the great local differences in rainfall. To obtain satisfactory results we must turn to some large area of uniform relief and geological structure, preferably some way from the sea and showing a steady progressive change of climate from south to north. Perhaps the best and most satisfactory example is to be found in European Russia. This great country stretches through some 30° of latitude, or over 2000 miles from north to south, and nearly as much from east to west; it is in the main one vast level plain from the shores of the Baltic to the Urals, and the geological structure is remarkably simple and uniform. Owing to causes which need not here be discussed the climatic belts do not run exactly east and west, but trend from south-west to north-east, and on the whole the soil-belts follow these very closely. The extreme north is exceedingly cold, in fact almost arctic in character, while in the south-east the rainfall is small and the summer climate hot. This part belongs rather to the arid zone, while the middle portion is more temperate, though everywhere the winters are cold.

From a detailed study of the soil-types of Russia Sibirtzev[1] has drawn up a zonal classification, and he finds that the arrangement of the soil-belts shows a rough parallelism to the climatic zones. In all six principal types can be recognized; besides these there are occasional patches of so-called interzonal soils, in places where local peculiarities of climate or of topography have given rise to well-marked special conditions. Again along the chain of the Urals and elsewhere are found certain superficial deposits, not true soils in the ordinary sense, but geological formations that may be considered independent of climate. The nature of these will be considered later.

The classification is based on genetic principles; that is to say that under identical climatic conditions rock-disintegration can efface original differences between rocks and gives rise to similar products. Hence a particular soil-type is the result of its geographical environment, that is, of the conditions dominant in its zone. In each zone there are local varieties but all possess certain characters in common.

Sibirtzev's classification, as applied to Russia, may be arranged as follows in a tabular form[2].

---

[1] Sibirtzev, "Étude des Sols de la Russie," Congrès Géologique Internationale St Petersburg, 1897, *Report*, p. 73. See also Glinka, *Die Typen der Bodenbildung*, Berlin, 1914 (a German translation of the author's Russian lectures, with a good coloured soil map of Russia).

[2] In the original publication the first type of zonal soils is laterite This however does not occur in Russia, and has here been omitted, to be treated along with other important types also absent from the list given by Sibirtzev.

A. Zonal soils.
  1. Aeolian soils.
  2. Dry steppe soils.
  3. Tchernozom or black earth.
  4. Grey forest soils.
  5. Podzols.
  6. Tundra soils.

B. Interzonal soils.
  1. Salt soils.
  2. Rédzina or Borovina soils.
  3. Marsh soils.

C. Azonal soils (or surface deposits).
  1. Screes, moraines, river gravels, sand dunes.
  2. Alluvium.

(1) Aeolian or dust soils. This group includes the soils that are formed from the Loess and other wind-transported deposits that are so abundant in the dry regions of the arid belt. (For an account of the origin of the Loess see p. 136.) There is usually little difference between soil and subsoil, chiefly owing to the small proportion of humus, which is usually less than 1 per cent., rarely rising to 2 or 3 per cent. The soils are commonly very fine in texture, somewhat loamy and generally of a yellow or even orange colour. When irrigated they are often fairly fertile, but commonly suffer from want of water, owing to the climate.

Soils of this class have a wide distribution in the Trans-Caspian region in Turkestan, and in many parts of central Asia and China. They are also found in the Great Basin region of the western and south-western United States. In the southern hemisphere the soils of the Karroo region of Cape Colony may also be referred here.

(2) Dry steppe soils (the chestnut-brown soils of Doku-tchaiev) occupy a considerable space in the south-eastern corner of European Russia, between the Ural and the Volga, along the shores of the Black Sea, and in parts of the Crimea. In this region the annual rainfall ranges from 7 to 10 inches

and vegetation is scanty. The subsoil is mainly soft Tertiary
or post-Tertiary clays, often with rock-salt and gypsum.
Patches of interzonal saline soils are common in depressions.
Towards the north there is a gradual transition to the black-
earth soils of the next zone.

With these may be compared the soils of the "Desiertos"
of eastern Spain, and some soils in California, Colorado and
New Mexico, as well as parts of the Orange Free State, and the
Transvaal in South Africa.

(3) The Black Earth or Tchernozom. This soil-type is
of enormous agricultural importance, constituting the great
wheat-growing area of Russia and other neighbouring countries
of south-eastern Europe. The black earth region mainly forms
one vast plain, with occasional ravines and hollows and in
these alone are trees found to any great extent. The rainfall
is small, from 10 to 13 inches only, and serious droughts are
frequent. Hence the climatic conditions are well marked and
have had an important influence on the formation of the soil.

The origin of the Tchernozom is a subject that has given
rise to much controversy and the literature of the subject is
very large[1]. From an exhaustive study of the characters and
distribution of the soil Dokutchaiev concludes that it always
shows a close genetic relationship with the underlying rocks,
and is not, as was formerly believed, a transported deposit of
marine or glacial origin. Tchernozom can be formed from
various rocks, especially Loess, Chalk, Jurassic clays, or
weathered granite. This affords a very clear illustration of the
general principle that under determinate conditions similar soils
may arise from very different rocks (i.e. the zonal principle of
Sibirtzev). It is not formed in tree-covered regions, but only
under grassy steppes. The soil has a very constant thickness,
rarely if ever exceeding 5 feet, and the structure is markedly
granular.

As its name implies Tchernozom is characterized by a very
dark colour, usually black, but occasionally shading into grey
or brown of a dark shade. Spots and concretions of calcium

---

[1] For a summary of the Russian publications on the subject, see Kossowitsch,
Die Schwarzerde, Vienna, 1912, a German translation of a Russian monograph.

and magnesium carbonates are common, often occurring in worm-burrows. The upper layers of the soil are exceedingly rich in humus, the amount sometimes rising to 16 per cent. and averaging about 10 per cent. Sometimes layers rich in humus are also observed deep down in the subsoil. These are no doubt analogous to the Ortstein of the northern soils (see p. 140). The humus appears to be derived wholly from the growth and decay of vegetation, especially the roots of grasses.

The following table from the work of Kossovitch, already cited, shows the chemical composition of two representative samples of Tchernozom, one rich in humus, the other rather poorer.

	Tchernozom, Bobrov, Gouv. Voronesh	Tchernozom, Nikolaievsk, Gouv. Samara
Humus	11·73	7·84
Water	2·94	2·87
Phosphorus pentoxide	·32	·28
Silica	59·93	64·40
Alumina	12·61	13·12
Ferric oxide	5·20	4·62
Lime	2·32	1·41
Magnesia	1·74	1·58
Potash	2·11	2·22
Soda	1·05	1·33
	99·95	99·89

The fertility of this soil is extraordinary; it appears to be able to grow heavy crops of wheat and maize for an indefinite time under an extensive system of farming, with little or no manure.

The general distribution of the Tchernozom in European Russia is simple; it forms a broad band stretching W.S.W. and E.N.E., from the western frontier of Russia, across the basins of the Dniepr, Don and Volga, towards the southern end of the Urals, where an interruption occurs. It covers also a wide extent in Siberia, in the Governments of Tobolsk and Tomsk, even occurring in patches as far as the basin of the Amur. The total area occupied by it in European Russia is estimated at about 350,000 square miles, and it also extends into Rumania[1], Galicia and Hungary.

With the Tchernozom may be compared the "black cotton soil" of India (p. 160), some of the prairie soils of North America (p. 152), and the black soil of parts of Argentina.

(4) Grey forest soils. The soils of this type form a narrow but fairly continuous band to the north of the Tchernozom, running from Lublin and Volhynia to Kama and Viatka. The district was probably once open steppe, but is now covered by forests. The soil lies on moraines and on argillaceous sediments. Though once doubtless rich in humus, this has been leached out by vegetation and now varies from 3 to 6 per cent. The soils are grey in colour and granular in structure. The thickness is small, rarely exceeding 10 inches; the subsoil is very similar and seldom more than a foot thick.

Grey forest soils are known to occur in Siberia, and they may also exist in Galicia, Hungary and in some German forests, as well as along the southern border of the Canadian forests, but little is known about them outside Russia.

(5) Podzol. This general designation is now commonly applied to a soil-type that covers enormous areas in European Russia, at least two-fifths of the entire country, as well as most of Siberia except the extreme north. The climate of this belt is fairly moist, with very cold winters and the country in its natural state is covered with coarse grass, brushwood, scattered trees and occasionally marshes. Beds of peat are fairly common. The podzols are in the main similar to the German bleached soils (Bleicherde), the principal characteristic being the leaching out of the soluble constituents from the upper layers, including the iron and phosphoric acid. These dissolved constituents are commonly reprecipitated some distance below the surface, forming a pan (Ortstein). The soils are white or grey in colour and very siliceous; they are very infertile unless well-manured, owing to the removal of plant food in solution. The subsoil is commonly of morainic origin.

Podzols are in the main sandy soils, very like the bleached

---

[1] Murgoci, "Die Bodenzonen Rumäniens," *C. R.* 1*ère Conf. Internat. Agrogéol.* Budapest, 1909.

sands so often found underlying the heather in central Europe
and in the east of England, wherever owing to small rainfall or
other causes the conditions are unfavourable for the growth of
peat. The name is more strictly applicable to the bleached
subsoil, as distinguished from the upper layer where humus is
more abundant. Where the latter layer is absent the land is
extraordinarily barren.

(6) Tundra. From the agricultural standpoint this can
scarcely be regarded as a true soil. The tundra is really a
gigantic peat bog which is permanently frozen at a depth greater
than 3 or 4 feet; its characters and origin are described in
Chapter v (p. 120). It is confined to the extreme north of
Russia, Siberia and North America and is entirely uncultivated.

Among the interzonal soils of Russia two or three types
only need brief mention. These are the saline soils of the
south-east, the Rédzina soils of Poland, and the marsh soils.

(1) The saline soils or Solontzy are found in the south-
eastern corner of European Russia in the Trans-Caspian territory
and in Turkestan. They also form occasional islands in the
Tchernozom belt. The soil is generally black above and grey
below, and at times, when evaporation has been active, the
surface is powdered with salt crystals. The subsoil is usually
a stiff brown or red saline clay. The salts present in the soil
are those usually associated with salt lakes, especially chlorides
and sulphates of sodium, magnesium and calcium. Since the
more important varieties of saline and alkaline soils are treated
elsewhere (see p. 156), it is unnecessary to pursue the subject
here.

(2) The Rédzina or Borovina soils of Poland. These soils
are of rather unusual character, since they are of a humous or
peaty nature, though lying on Chalk or limestone. The uppermost layer is dark grey in colour, owing to the humus; then
comes usually a layer of clay or marl with the unaltered calcareous rock below. The origin and affinities of these soils
have not yet been sufficiently explained, but their analogues
might be sought in some of the East Anglian heaths lying on
Chalk.

(3) Swamp soils. These soils are almost exactly similar

to the fen peat of England and Germany; according to the published accounts, the plants that take part in their formation belong mainly to the same species of rushes, sedges, grasses, *Butomus, Sagittaria*, etc., as are found in the English Fenland. A detailed description is obviously unnecessary (see also p. 119).

The azonal soils of Sibirtzev's classification are in the main surface deposits, which can scarcely be reckoned as soils in the ordinary sense. They are found in regions where special local conditions have given rise to extensive superficial deposits of various kinds, largely independent of climate and of geological structure. Here for example may be included the screes of mountain ranges and of stony deserts, boulders of disintegration, especially those lying on outcrops of igneous rock, bare rock surfaces of any kind, including some limestone plateaux with underground drainage, the gravel beds of rivers, the sand-hills of the sea-shore, the moraines, eskers, drumlins, gravels and sands of glaciated regions and others. In this class also may be placed the alluvial deposits of rivers and deltas, and the lagoon deposits of the sea-shore. All of these when weathered give rise to more or less soil, those of the alluvial class often being extremely fertile. The origin of all these deposits is treated in Chapter v. It is evident that the soils, if any, derived from them will show great variety, owing to local conditions, varying with the lithological character of the deposits themselves, so that no general account can be given.

Soil-zones in North America. As already remarked the majority of, if not all, the Russian soil-types here enumerated can be identified in North America. Here owing to the disposition of the climatic zones, the soil-belts trend from south-east to north-west, the hottest and driest region being in Arizona, Nevada, Utah and southern California, where the average temperature for the month of July exceeds 90° F. Here are found representatives of the saline and dry steppe soils. Aeolian soils, Loess or Adobe are common in Colorado, Utah and Wyoming, while the black prairie soil, the representative of the Tchernozom, covers an enormous area in Montana, Dakota, Nebraska, Iowa, Wisconsin, Minnesota, Kansas, Missouri, Arkansas, and Oklahoma. It also extends widely

in the corn-growing regions of Canada, in Manitoba, Saskatche-
wan and Alberta. Grey forest soils exist in the northern part
of the American prairies, bordering on the forest zone. In
the latter, soils of the podzol type have a wide extension, while
in the far north the tundra region is very extensive in Alaska
and east of the Mackenzie river. Thus it appears that in western
America the zonal distribution of soils is quite as strongly
marked as in Russia. In eastern America however the soils
appear to show more resemblance to those of western Europe.

Although the soils of Russia and western America afford
perhaps the best examples of zonal arrangement, nevertheless
several of the most important and widespread types are there
unrepresented; some notable examples are the lateritic and red
soils of the tropical and subtropical regions and above all the
ordinary arable and meadow soils of western Europe and the
eastern states of America. These and others it is now necessary
to treat in some detail.

Lateritic soils. The origin of laterite has already been
discussed (pp. 57 and 116). It was there pointed out that laterite
results from a special kind of chemical weathering, the dominant
feature being a hydrolytic decomposition of silicates, with
great loss of free silica and of alkalies, lime and magnesia, and
a consequent increase by concentration of iron and alumina.
This process is if anything still more accentuated in the superficial layers, and owing to the generally abundant rainfall of
the tropics there is much solution and oxidation and little
accumulation of humus. Laterite soils are therefore generally
red in colour, and are often strikingly deficient in some of the
elements of plant food. Laterite formation is clearly determined by climate, and appears to be best developed where
there is a marked alternation of wet and dry seasons. According
to all authorities laterite formation is the distinctive and
characteristic geological process of the tropical zone.

The principal constituents of laterite soils are hydrated
oxides or hydrates of iron and alumina, together with a considerable amount of unweathered quartz. In India especially
calcareous concretions are often found in abundance at some
distance below the surface; these are known locally as *kankar*;

part of the iron oxide is often also in a nodular and concretionary form.

Red soils. Highly characteristic of the subtropical and
warm temperate regions is a group of soils distinguished by a
prevailing bright red tint. These are not sharply marked off
from the laterite soils of the tropics on the one hand or from the
brown meadow soils of the cooler temperate belt on the other,
but they undoubtedly form a natural class of soils whose
characteristics are mainly determined by climatic conditions.
Red soils are most perfectly developed in regions with a hot
summer and a fairly warm winter; they appear to belong to
the northern part of the arid belt, where a fair amount of rain
falls, at any rate in the cooler part of the year, though the
summer is dry. In this they show an affinity to the lateritic
group. The most favourable conditions for their development
are realized over a great part of the Mediterranean region, in
parts of Spain, Italy and Greece, in Corsica, Sardinia, the
Balearic Islands, and over a large tract in northern Africa and
Asia Minor. A special variety is the terra rossa of the Dalmatian
coast of the Adriatic[1].

Ordinary red soils show great variety of constitution,
ranging from sands to clays, and it appears that the characteristic bright red colour is conditioned by two circumstances,
namely, low content of humus, and abundance of ferric oxide;
these conditions are evidently determined mainly by climate;
during the mild and fairly moist winter humus is rapidly
destroyed, and in the dry season there is much precipitation of
iron oxide from the evaporating solutions in the soil.

Ordinary red soils may apparently be formed under suitable
conditions from rocks of almost any kind, but terra rossa in
the strict sense is formed especially from limestones that
originally contained a certain amount of iron and alumina.
Terra rossa is best seen in the Karst district, in the south-west
of the Austrian empire, bordering the Adriatic. This region
consists mainly of limestones and dolomites of Triassic age,
and shows in a high degree of perfection all the special features

[1] Neumayr, *Verhandl. Geol. Reichsanstalt*, Vienna, 1875, p. 50. See also
Erdgeschichte, Vienna, 1895, p. 453.

of limestone denudation. Terra rossa is a very ferruginous clay that is supposed to represent the insoluble residue of the impure limestones; it is therefore possibly comparable with parts of the English clay-with-flints (see p. 114). However it has been suggested that the red material is really due to precipitation of iron oxide from solutions of iron salts and not directly to normal weathering. This question is still undecided. It is noticeable that red soils extend further north on limestones and dolomites than on any other sort of rock and this fact favours the "insoluble residue" theory; also when specimens of the underlying white limestones are dissolved in weak acetic acid, a small amount of red ferruginous clay is left behind, hence the argillaceous material is really present in the rock.

Brown soils. Under this general designation are here included the ordinary arable and pasture soils of western and central Europe, the eastern United States and many other regions of cool temperate climates. Even in the areas thus specified the annual range of temperature varies a good deal, being least in the insular climate of the British Isles and other maritime regions, increasing towards the interiors of the continents. The chief determining factor seems to be a fairly abundant but not excessive rainfall, and a range of temperature that admits of the free growth of herbaceous plants, without too strong a tendency towards forests or peat-formation. The brown soils are in fact *par excellence* the soils of fertile and highly cultivated lowlands and plains, especially adapted for arable land and rich permanent pasture. Under present economic conditions they are mainly cultivated on the intensive system, with liberal manuring and heavy crops in regular rotation.

Owing to abundant and free water-circulation much soluble material is washed out of the upper layers, especially sulphates and carbonates, whereas alumina and phosphoric acid tend to accumulate. In many soils of this class potash and lime tend to be deficient, especially the former.

The colour of the soil depends on a combination of two factors: (1) the relative abundance and state of oxidation of the iron, and (2) the proportion of humus. The latter can be removed from a sample by treatment first with very dilute hydrochloric acid and then with ammonia, the true colour of the soil thus becoming apparent. This is found to depend mainly on the character of the rock or other deposit from which the soil is derived. The colour is usually some shade of brown, yellowish or reddish, according to circumstances. From red rocks, such as the Old or New Red Sandstone, red soils are formed. When the content of humus is unusually high the soil may be nearly black, while Chalk soils are often very pale in colour. However the calcareous soils in general form a somewhat aberrant group.

Brown soils show the utmost variation of agricultural character, ranging from the heaviest clays to light sands, and include all the varieties known as loams and marls. The mineral composition and origin of soils has already been treated at some length, and in the later chapters of this book a detailed account is given of the dominant agricultural characters of the soils derived from the rock-formations of the British Isles. It is therefore unnecessary to pursue the subject further here.

Saline soils. Under this heading are included two very different types of soil, agreeing only in the fact that both are unusually rich in soluble mineral substances. The first group includes the soils of those areas which are occasionally overflowed by the sea, or into which sea-water is able to penetrate by percolation. They are necessarily confined strictly to coastal regions at or near sea-level. Such are the salt marshes of the Thames estuary, of parts of the Norfolk coast and the "slob lands" of Ireland. The soil of such areas is usually alluvial in origin, being either sea or river silt and commonly of a muddy nature. The salts present are naturally those found in sea-water, the most important being sodium chloride, with a smaller proportion of magnesium chloride and sulphate. When capable of utilization for agricultural purposes such areas are always pasture, and the herbage is of a peculiar character, though often forming good feeding for stock.

The second type of saline soils belongs exclusively to the arid climatic belts and occurs in those regions where the drainage is deficient. In all soils soluble salts are formed by chemical weathering; in moist climates these are leached out by rain and carried away in solution in the drainage water, eventually reaching the sea, adding to its stock of salt. But in regions of deficient rainfall this leaching action is in abeyance, the soluble salts tending to accumulate in the soil, till they may become conspicuous, as in the salt deserts of south-western Asia. Where the drainage basin is entirely self-contained, without outlet to the sea, the effect of such leaching and stream-action as there may be is to concentrate the salts in the hollows of the surface. The extreme form of this is seen in the formation of salt lakes, such as the Dead Sea and the Great Salt Lake of Utah, both of which were once fresh, but have now become many times salter than the sea by concentration and evaporation. In some cases it has been shown that the saltness of the soil is due to the ascent of saline solutions by capillary action from lower levels, where beds of soluble salts of an earlier geological date have been covered and preserved by later sediments. This appears to be the case with the Szek lands of Hungary and some of the salt soils of western America, and the extreme saltness of the Dead Sea is attributed to solution by the Jordan waters of beds of rock salt and other saline substances in the Cretaceous and Tertiary strata of the district.

The composition of the salts found in such soils varies widely according to circumstances. The commonest are magnesium chloride, magnesium sulphate and sodium chloride; potassium compounds are less abundant, though almost always present. Calcium sulphate is also very widely distributed. Soils rich in the constituents just mentioned form the true saline soils in the strict sense, while those characterized by sodium sulphate and sodium carbonate are generally called alkali soils, though some American writers use the term alkali soil indiscriminately for both groups. German writers, on the other hand, use the term saline soils (Salzböden) to include soils with sodium sulphate and carbonate as well as those with chlorides.

Several different types of saline soils have been recognized by Glinka and others[1]. They are closely related to the ordinary

[1] Glinka, *Die Typen der Bodenbildung*, Berlin, 1914, pp. 177–211.

soil-types of arid regions, being generally poor in humus and pale in colour, showing much resemblance to the bleached soils of the German and Russian classifications. Some occurrences of this type in hollows in the steppe regions are described as "saline podzol" or as "solonetz" (see p. 151). Some varieties shrink and crack deeply when dry, developing a curious columnar structure, like miniature basaltic columns. In the neighbourhood of the Caspian Sea there is a considerable area of white soil or Byelozom, white or grey in colour and rich in salts. This is characteristic of the southern dry steppes, where the annual rainfall is about 8 inches.

Alkali soils. This term is unfortunately used in a somewhat vague way, especially by American writers, to include a considerable variety of soils characterized by an excessive proportion of soluble salts. Two chief types are generally recognized in America, namely, black alkali soils, characterized by sodium carbonate, and white alkali soils, in which sodium sulphate is the dominant soluble constituent. The term "alkali soil" should strictly be reserved for the first class only. Some writers go even further and include in this group all the soils with a large proportion of chlorides, or what in Europe are commonly designated saline soils.

The alkali soils are, as would be expected, restricted to the arid regions of the globe; they are found extensively in western America, northern Africa, and south-western Asia, as well as to a more limited extent in India. They probably occur also in South Africa and Australia, though as to this we possess little information. In Europe they occur in the steppe region of south-east Russia, and to a small extent in the Szek lands of Hungary, along the course of the river Theiss. Both black and white alkali soils are represented here, although this is not an arid region, and the salts may be derived from underground salt beds[1].

Normally alkali soils are found only in regions of deficient rainfall, where there is no surface drainage to carry away the soluble salts from the soil, and the only loss of water is by

---

[1] Glinka, *Die Typen der Bodenbildung*, Berlin, 1914, pp. 180–182.

evaporation. Under these conditions the salts tend to accumulate in the upper layers of the soil, commonly in the uppermost three or four feet. The nature of the salts will of course depend on the source from whence they are derived, that is, the salt-content of the underlying rocks. Whereas in ordinary saline soils the chief salts are sodium and magnesium chlorides, in alkali soils the most important constituents are sodium sulphate (Glauber's salt) and sodium carbonate. Soils rich in the latter salt can generally be distinguished by the occurrences of surface puddles of black or dark brown water; hence the name of black alkali soil applied to this class. The black colour is supposed to be due to humus dissolved in the salt solution.

The distribution of soluble matter in alkali soils varies widely according to the meteorological conditions prevailing at the time. During rainy weather the salts are dissolved away from the surface and carried downwards. When the air is dry, on the other hand, evaporation takes place from the surface and the salts accumulate in the upper layers of the soil. In extreme cases an incrustation of salt crystals may even form on the surface, as in the alkali deserts of western America and many other places.

Alkali soils are usually very infertile, partly owing to the deleterious effect of the strong salt solutions on plants, and partly owing to the peculiar effect of alkaline salts in forming a hard pan in the upper part of the soil (see p. 142). The native vegetation is scanty and peculiar, such as the Australian "salt-bush" and the American "sage-brush," most of the plants not being palatable to stock, while few ordinary agricultural crops will succeed on such soils. True alkali soils are therefore generally barren and uncultivated, though often rich in the elements of plant food, with the important exception of nitrogen, which is always scarce. When they can be irrigated sufficiently to wash the harmful salts out of the soil they may become quite fertile[1].

The soil-regions of India. The soils of India have been investigated by several authorities, especially by Voelcker,

[1] Hilgard, *Soils*, New York.

Leather and Mann. Their general conclusions may be summarized as follows; there are four main types, distinguished
by origin and composition; within each type varieties can be
recognized. The four chief groups are:

    (1) The alluvium of the Indus, Ganges and Bramaputra.
    (2) The black cotton soil or regur of the Deccan plateau.
    (3) The red soils of Madras.
    (4) The laterite soils.

    (1) The alluvium of northern India varies greatly in its
character, according to climate and rainfall. In northern
India the rainfall is deficient in the west and increases steadily
eastwards. Parts of the Punjab and Rajputana are very arid,
whereas in the Khasia Hills, in the basin of the Bramaputra,
is the highest rainfall of the world (600 inches per annum).
Hence the soils undergo a very varied amount of leaching, this
being the determining factor. The alluvium consists entirely
of fine sand and silt, any larger elements being rare or absent.
Calcareous concretions (kankar) are abundant in the drier
regions. In the west the soils are of a distinctly arid type, and
in Aligarh and the Agra district even alkali soils (reh soils) are
found. Over the whole drier part of this belt, owing to absence
of excessive leaching, the soils are rich in potash, lime and
magnesia. On the other hand, in the Bramaputra basin
leaching is excessive owing to the heavy rainfall, and the soils
are very poor in soluble matter, especially lime. They are
consequently found suitable for the growth of tea.

    (2) The regur or black cotton soil forms a very well-marked
and distinctive type, having a wide distribution over the
plateau of the Central Provinces, the Deccan, from the Vindhya
mountains southwards. The soil is very deep, and possesses a
uniform black colour, even at a considerable distance from the
surface. Kankar concretions are common. In the dry season
the soil shrinks and forms very deep and wide cracks. In
regard to chemical composition, the percentages of potash, lime
and magnesia are high, but the nitrogen is curiously low,
considering the notable fertility of the soil, which is said to
have been cultivated continuously for over 2000 years without

manure. It can only be supposed that nitrification is active, and that a supply of nitrate sufficient for the needs of the crop is constantly being produced and removed as fast as it is formed. The origin of this soil is somewhat of a mystery; it lies mostly on basalt, but scarcely possesses the characters that would be expected of a sedentary soil on that rock. On the other hand any theory of formation by transport also presents considerable difficulties.

(3) The red soils of Madras are clearly of sedentary origin and are mainly formed from metamorphic rocks, but they are also known on other substrata. They are very variable in their chemical composition, agreeing only in the prevailing red tint, due to the presence of finely disseminated ferric oxide. The total percentage of iron is not as a rule notably high, and the effectiveness of this substance as a colouring matter must be attributed largely to its fine state of division. The red soils are as a rule less rich in plant food and less fertile than the black cotton soil. This soil type is in no way remarkable and must be regarded merely as an instance of the red soils so characteristic of tropical and semi-tropical regions; a general description of this type has already been given.

Chemical Composition of Indian Soils.

	Alluvium of Ganges, Punjab	Alluvium of Bra- maputra, Bengal	Regur, Trichi- nopoli	Red Soil, Madras	Laterite Soil, Madras
Insoluble matter	81·57	84·60	65·16	90·47	76·86
Potash ...	·74	·35	·14	·24	·0)
Soda	·08	·30	·01	·12	·17
Lime	1·44	·04	2·18	·56	tr.
Magnesia ...	1·97	·46	2·47	·70	·77
Ferric oxide ...	4·32	2·03	9·27	3·51	10·09
Alumina ...	5·85	5·03	13·76	2·92	8·84
Phosphoric acid	·23	·05	tr.	·09	tr.
Carbon dioxide	1·13	—	·91	·30	·12
Water and organic matter	2·67	5·59	5·85	1·01	2·87
Total	100·00	98·50	99·75	99·92	99·81

(4) The laterite soils occur chiefly in the coast region of Bengal, in Madras and on the Malabar coast. They appear to

grade into the red soils in places, and need no further description here. (For a general account of such soils see p. 154, and for the origin of laterite p. 115.)

The table on p. 161 shows analyses of a typical example of each of the types mentioned above.

General classification of climatic soil-zones. From a study of the work on the geographical distribution of soil-types carried out by many investigators in Germany, Russia and America, certain general conclusions can be drawn. In the first place it is clear that in soil-formation the dominant controlling factor is climatic rather than geological; that is to say, that under determinate conditions of temperature and rainfall, rocks and superficial deposits of the most varied kind may give rise to almost identical soil-types. Of these climatic conditions the most potent is undoubtedly the rainfall; temperature plays a secondary part, and its action is indirect rather than direct; temperature is certainly of great importance in controlling the growth of vegetation, both quantitatively and qualitatively. It is largely through vegetation (or the want of it) that the special characters of each soil are developed.

So far as the rainfall is concerned, its absolute amount is the chief consideration, but it must not be forgotten that the *relative* moisture-content of the atmosphere also has much influence; in a very hot region much more water vapour is required to saturate the atmosphere than in a cold one, hence with different annual average temperatures the same rainfall may produce dry air in one case (that is, air far below its saturation point), and in the other case air completely saturated with moisture. The effect on vegetation will obviously be very variable; one region may be inhabited by a swamp-flora, the other by xerophytic plants.

Even, however, when this cause of variation has been brought into account, it appears that the total rainfall is the most important controlling influence in soil-formation. In a dry region chemical decomposition is in abeyance; rock-disintegration is mainly mechanical, and wind-transported deposits are the rule. Increased rainfall gives rise to more active chemical and bacterial processes, and water-transport is dominant. In high northern latitudes the conditions are special, owing to the more or less permanent freezing of the subsoil; this leads to the development of the tundra type, which is independent of rainfall.

On a climatic basis the earth can be divided into soil-zones, which are of general application, as follows[1]:

Annual rainfall			Soil type
0- 8 inches	...	...	Desert.
8–16 ,,	...	...	Steppe.
16–24 ,,	...	...	Prairie.
over 24 ,,	...	...	Forest.
Arctic temperature	...	...	Tundra.

These zones are only applicable with strictness to regions of the world, such as Russia, North and South America and East Africa, which are still in a more or less natural state. In western Europe, India, China and other lands of ancient civilization there has been so much interference by the hand of man, through clearing of forests, arable cultivation, irrigation and other processes, that the original character of the soil has been profoundly modified. It appears however that the greater part of the British Isles, France and southern and western Germany originally belonged to the forest type, as also does the eastern part of the United States and of Canada. Tropical forests are chiefly found in South America, Central Africa and parts of southern Asia and the adjacent islands. As European examples of the prairie type we may instance Hungary and central Russia; here also must be classed the prairies of North America, the Savannas and Llanos of South America, the grassy park-like plains of East Africa, and the pastoral districts of Australia. Steppe conditions are found largely developed in Russia and Central Asia, Spain, South Africa, the south-western United States and parts of Australia. It is unnecessary here to enumerate the desert regions of the world, which are widely spread both in the northern and southern hemispheres; they are of no agricultural importance, though of great interest to the geologist. It is possible, however, that in the future,

---

[1] Von Cholnoky, "Über die für Klimazonen bezeichnenden Bödenarten," *Comptes Rendus* 1ère *Conf. Internat. Agrogéol.* Budapest, 1909, pp. 163–176.

by means of irrigation, large areas now desert may be brought under profitable cultivation.

Soil types in Britain. When we come to consider in detail the soils of any particular area we shall find wide apparent variations in their characters, especially in regions of such variable geological structure as the British Isles. That this is the case will appear from the chapters devoted to the consideration of the stratified and other rocks of this country (Chapters IX to XVI). Nevertheless when climatic factors are also taken into account, it appears possible to distinguish some broadly defined regions in which certain similarities run through soils of the most varied origins. These broad variations depend largely on temperature, rainfall, relief of the ground and vegetation, the last being largely controlled by the other three. By a careful comparison of British soil types with those previously discussed as characteristic of certain climatic zones in Europe and America it will doubtless in time become possible to assign each region to its proper class. As yet sufficient materials are not available for detailed statements of this kind, but even now a small attempt may be made on general lines. The differences mainly depend on the extent to which soluble salts have been washed out of the soil by percolating water, which is evidently a matter of rainfall. As would naturally be expected there are marked differences in this respect between the eastern and the western sides of the country, the soils on the west being more washed than those on the east. In Devon and Cornwall and in the south-west of Ireland this effect of rainfall is partly compensated by the higher temperature, which causes a greater luxuriance of plant growth. In the Highlands of Scotland and to a less extent in the north-west of Ireland, lower temperature and greater elevation are specially favourable to the growth of peat. Many of the soils here consist almost entirely of sand and humus, all soluble matter having been washed away. In the western half of England the loss of soluble matter is greater than in the east, and the soils are often deficient in some of the elements of plant food. In the dry regions of the east soluble plant foods tend to concentrate in the soil owing to the smaller amount of percolating water, while humus is on the whole less abundant. (Here such exceptional cases as the Fenland, due to special causes, must be eliminated.) On the whole the soils of the west of the British Isles seem to approximate to the Podzol type of Russian authors, while those of the eastern parts may be regarded as typical "brown soils" of the continental classification. Nevertheless there are many local variations and the above statement must be regarded as merely an approximation to the truth. The detailed study of the soils of an area of sufficiently varied structure and relief would certainly lead to interesting results from this point of view; it may perhaps be suggested that Yorkshire offers a promising field for investigation. The four chief regions, the western hills, the central plain, the Cleveland Hills and the Wolds, vary so much in altitude, climate and rock constitution that many of the chief soil types of continental authors should be here recognizable. At any rate, when regarded from the conventional standpoint, the county seems to contain soils of almost every conceivable kind.

CHAPTER VII

THE GEOLOGY OF WATER SUPPLY AND DRAINAGE

One of the most important agricultural applications of geology is in connexion with water supply. It is seldom that a farm has the advantage of access to a public water supply; even if such is available for the house and buildings, the live stock in the fields in general have to depend on ponds, springs, wells and streams; in other words on the natural water supply of the land. Every field that is likely to be pastured by stock should have a sufficient supply of pure water, preferably running water, but seldom is this state of things attained or even possible of attainment.

Water supply is entirely a geological question; the primary source of all is of course rainfall, and in this sense it is a matter of climate; but given a sufficient rainfall, the availability of this for practical purposes depends on the nature and distribution of the rocks of the district, together with the superficial deposits that in most places overlie the hard rocks, these being often of great importance as controlling factors.

Rainfall. As is well known to every one, rainfall varies much from place to place. In some regions, such as the coast of Peru, parts of the Sahara and the interior of Australia, there is none at all. Over wide areas of the world the climate may be described as dry or arid, leading to the formation of deserts. In these the annual rainfall is less than 10 inches. When the rainfall exceeds this figure, or even in favourable localities when it falls considerably below this limit, some sort of agriculture is possible, especially the keeping of sheep and goats. Here, of course, the possibility of artificial irrigation is for the present left out of account. Over the greater part of the temperate zone the annual rainfall exceeds 20 inches, while in parts it is far higher than this. Even within the limits of the British Isles there is great variation. In the south-eastern counties the average rainfall is about 24 inches; to the north and west of a line drawn from the mouth of the Tees to Bristol it nearly always exceeds 30 inches, while at special places it is far greater; at Seathwaite in Cumberland the annual average is about 130 inches, while here, as also near Snowdon and at Ben Nevis, records of 220 to 230 inches per annum have been obtained. The highest rainfall of the world, so far as known, is found in the Khasia Hills in northern Bengal, in the foothills of the Himalayas, where the annual average is about 600 inches, nearly all of which falls in about three months. This is of course due to exceptional conditions, namely, the occurrence of a regular monsoon, with well-marked wet and dry seasons.

Of the total amount of water that falls as rain, it has been estimated that one-third runs off at once as streams, one-third sinks into the ground to form underground water and one-third passes back into the air by evaporation. It is with the first two portions that we are here concerned. It is unnecessary however to say much about surface streams; the origin and development of river systems has already been described in Chapter III, and it is obvious that a river or stream can be utilized in many ways for agricultural purposes. Live stock can obtain their drinking water directly from the river; water can be pumped from the river into artificial ponds and ditches, and thus conducted from place to place; under favourable conditions water power can be used to drive agricultural machinery and so on. Besides this, there is the very important subject of irrigation. In Britain this process is not much employed, being confined chiefly to the water meadows of the south-western counties and to sewage farms. In dry regions however, such as South Africa, Australia and parts of western America, irrigation is of enormous and increasing importance. This, however, is engineering rather than geology and the subject cannot here be pursued further.

Underground water. The geology of water supply deals mainly with the second fraction as above described, namely,

the water that sinks into the ground.   The ultimate fate of this
must obviously depend on the geological structure of the area
and on the nature of the rocks composing it.   The most
important consideration here is evidently the permeability or
otherwise of the rocks.   If they are totally impermeable,
water cannot sink into the ground at all; it must run off as
streams or be evaporated into the atmosphere, or it may, of
course, remain on the surface as ponds, lakes and swamps, if
the configuration is favourable.   Ordinarily the rocks are
more or less permeable and the water works downwards,
forming what is known as *ground water*.   Experience in deep
mines shows, however, that this water has not an indefinite
downward extension; below a certain depth, depending on
local conditions, all mines are dry.   There is one exception to
this rule; in certain volcanic districts hot water comes up
in large quantity from a great depth within the crust, in the
form of geysers and hot springs.   This water is evidently of
different origin from the atmospheric water previously considered
and is believed to be derived directly from the heated igneous
material of the interior of the earth.   It is of no agricultural
importance and need not be considered further.

Under ordinary circumstances, then, a certain zone of the
earth's crust is saturated with water of atmospheric origin,
which is often called meteoric water; the lower limit of this is
indefinite, but the upper limit is much better defined and is
often called the *water table*.   This surface however is not by
any means horizontal, and in a general way it often follows the
undulations of the land, maintaining a sort of rough parallelism.
In places the water table intersects the land surface and gives
rise to springs, which often tend to occur in lines.   The position
of the water table is very largely controlled by the degree of
porosity of the rocks and their capacity on the one hand for
holding water or on the other hand for preventing its passage.

Pervious and impervious rocks.   The amount of water that
can be contained by any given rock, when saturated, depends
on its texture, that is to say chiefly on the size of the particles
and the amount of open space between them.   Loose and incoherent deposits, like gravel and sand, will hold more water than

similar rocks that have been welded into a solid mass by deposition of some cementing material in the pores between the particles. Another consideration of very great importance is the presence or otherwise in the rocks of open joints, which may facilitate the passage of water through even the most naturally impervious rocks, and to a certain extent may even permit storage in them. Even the most impervious rocks will absorb and hold a certain amount of water, but if the grain of the rock is very fine, internal friction will prevent the movement of this water. On the other hand water flows freely through rocks with plenty of pore-space, where the internal friction is small.

The following table gives a classification of the common rock-types according to their behaviour towards ground water, under three headings:

I. Porous and permeable rocks	II. Rocks holding water in fissures	III. Impervious rocks	
Sand	Quartzite	Clay	
Sandstone	Grit	Shale	
Gravel	Conglomerâte	Marl	
Sandy limestone	Marble	Brickearth	
Chalk	Slate	Marble	when not fissured
Oolite	Granite	Schist	
Dolomite	Greenstone	Granite	
Brown ironstone	Gneiss	Greenstone	
	Schist		

Springs. When rain falls on a surface of porous soil or rock it sinks in and works its way downwards until it comes to some obstruction, commonly a mass of impervious rock. Its further course will depend on the form and disposition of this barrier.

Fig. 34. Formation of springs in a hill of horizontal strata, pervious above, impervious below. The broken line indicates the upper limit of saturation (water table). S, S = springs.

The simplest case is where a hill consists of horizontal strata, pervious above, impervious below, as shown in the figure.

Then the water will come out as a line of springs or of ill-defined
soakage, along the junction of the two rocks; in this case all
round the hill.   If the strata are inclined the water will of course
tend to run more freely down the dip, but owing to internal

Fig. 35.   Formation of springs in a hill of inclined strata.
S = principal spring, s = subsidiary spring.

friction there may be springs on the upper side also.   Theoreti-
cally the water might issue at any point along the plane of
contact, but in practice, owing to slight local inequalities which
tend to become accentuated in course of time, there are usually
more or less well defined springs at certain points.   These
form streams and cut gullies on the hills, thus helping to keep
the springs to a fixed point of discharge.   There are many
possible arrangements of rocks that can give rise to springs,
but the general principle is the same in all cases;   the water
percolates through permeable rocks and is stored in them,
until the rock can hold no more, when it overflows as a spring.

Owing to special local causes springs may be thrown out at
various points.   Dykes of igneous rock cutting across the

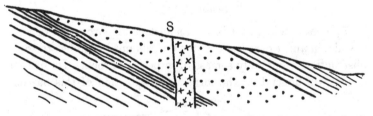

Fig. 36.   A vertical dyke of impervious igneous rock, cutting across inclined
strata, causes a spring at S by holding up the water in the pervious (dotted)
stratum to the left.

strata are generally impervious and may originate an accumulation of water, with outflow of a spring, in the middle of a
permeable series (see Fig. 36).   Again faulting may give rise
to favourable conditions, by bringing impervious strata against
permeable ones, and so forming a basin.   The conditions here

may somewhat resemble those necessary for an artesian well, to be presently described. In point of fact the possible variations are endless, and no good purpose would be served by a further detailed consideration of them.

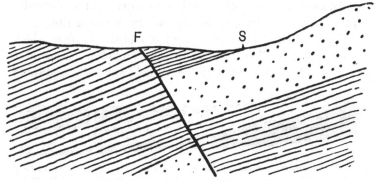

Fig. 37. Spring due to faulting of inclined strata. The permeable (dotted) bed is brought by the fault F against an impervious bed, and water is held up in it, till it rises to the point S, where it overflows as a spring.

Wells. The essential feature of a well is that it must cut the water table or upper limit of saturation. If this condition is fulfilled the portion of the well below this level will be permanently filled with water. The conditions may even be such that the water rises to the surface, or even above it as a jet. Such are called flowing or artesian wells. More commonly however the water has to be brought to the surface by a bucket and rope, or by a pump. If the level of the water is not more than about 30 feet from the surface the common suction pump may be used; beyond this depth it must either be brought up laboriously by dipping and winding, or by some sort of force pump, or ram. For this purpose windmills are very commonly employed, being both cheap and effective.

In choosing the position for a well it is essential to pay attention to geological structure. A shallow well sunk in a porous superficial stratum, such as gravel, will generally yield a supply of water which percolates from the surface. Such water is specially liable to contamination and is unsuited to domestic use, or for dairy purposes, though it may be employed in stables or for field troughs. The danger is somewhat

decreased if the well is lined with water-tight cemented brick-
work to as great a depth as possible, so that water can only
enter after being filtered through a considerable depth of
gravel or sand. In all cases special care should be taken that
surface water does not run directly into a well used for domestic
supply. This can be guarded against by a raised rim.

The most satisfactory wells are undoubtedly those in which
the water has to filter itself by passing through a considerable
thickness of porous strata in some such way as is shown in

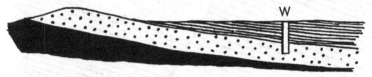

Fig. 38. Conditions favourable for a well (*W*). The water-bearing stratum
    (dotted) lies between two impervious beds. The rain falling on the higher
    ground to the left runs down the dip and is tapped by the well; it should
    rise to a considerable height above the top of the dotted bed, possibly
    even to the surface.

Fig. 38. Here the risk of contamination is small, the water
being derived from an upland region, and taken at a lower
level after percolating through a good thickness of sand. If
however there are buildings situated on the outcrop of the sand
(Fig. 39) there may be danger even in this instance.

Fig. 39. Diagram to show how the water of a well may be in danger of pollu-
    tion from a village lying on the water-bearing stratum further up the
    dip-slope. *W* = well sunk through impervious rock to pervious stratum.

The best of all water supply is that obtained from deep
artesian borings, such as those that penetrate into the Chalk
in various parts of London. The general principle involved is
shown in Fig. 40. Here a pervious stratum lies between two
beds of clay, the whole series being bent into a trough or syncline.
Rain falls on the outcrop of the pervious bed and runs down
the dip towards the centre of the trough; here it is under

pressure of a considerable head of water, owing to the difference
of level, and if a boring is made through the upper bed of clay,
the water will rise in it and may form a jet rising above the
surface. This happened when the first deep borings were made
in the London basin, but now owing to the number of such deep
wells, the pressure is much reduced and the water generally has

Fig. 40.   Conditions for artesian well.   A water-bearing stratum lies between
two impervious ones, the whole being folded into a trough or syncline.
The water falling on the hills at either end runs down the dip and collects
at the bottom; when the upper bed is pierced by the well the water is
forced up by the pressure of water above and may form a jet above the
surface.

to be pumped to the surface from some depth. As shown in
Fig. 37, a somewhat similar effect may be produced by a fault,
bringing a thick bed of impervious rock against a pervious
inclined stratum.   The name artesian well is derived from the
province of Artois in the north of France, where they were
first employed.

Drainage. A large proportion of the agricultural soils of
the British Isles contain normally more water than is beneficial
for crops; the object of land drainage is to ameliorate this
condition of the soil and to promote the growth of crops by
getting rid of the hurtful excess of water.   Certain large tracts
of land are water-logged simply by reason of their low elevation;
whatever the nature of the soil and of the underlying rocks the
water cannot run off, if there is no fall, as for instance in the
Fenland of eastern England and some of the "moors" of
Somerset. Here pumping is obviously essential, as well as
drainage, and the whole matter becomes one of engineering
on a large scale.   In this place however attention must be
confined to the geological aspect of ordinary farm drainage
where sufficient fall is available to carry off the water, if it
can be got into pipes or open culverts.

Rocks have already been classified into pervious and
impervious: as a first general rule it may be laid down that

soils lying on and derived from the impervious rocks require artificial drainage, while the others do not.   There are, of course, exceptions to both these statements, due to particular causes, but the general principle holds good.   The presence or absence of open joints in the rocks below is evidently an important factor.   Many high-lying and moorland areas, though lying on light soils, are actually in need of drainage from a combination of causes; in the first place the cool damp climate favours the formation of peat (see p. 118), and this in itself indicates a water-logged condition of the soil.   Again in such places impervious "pans" are often formed (see p. 140), thus holding water in the upper layers of the soil.   If the pan can be broken up by mechanical means, e.g. steam ploughing, the necessity for pipe-draining may be obviated.

The simplest type of drainage, often employed in upland regions, is that called "sheep draining"; this consists in cutting a network of surface channels or shallow gutters in suitable directions to carry off the rain-water before it can sink into the ground.   This can evidently only be effective on sloping ground, where the run-off will be fairly quick, and the channels must cross the slopes obliquely.   Surface drains and open drains of all kinds can only be makeshifts in agricultural lands, though deep open drains may sometimes be profitably employed in plantations of timber trees, where the existence of deep holes will do no harm to man or beast.   In nearly all cases where draining is necessary at all, it must be effected by means of pipes, or tiles as they are often called technically.

It is usually considered that to render a soil capable of growing its best crops, the upper surface of the ground water (water table) should be kept at least 4 feet below the surface. This does not necessarily mean however that the tiles must always be at that depth.   Owing to capillarity water will rise into drains for some distance above the water surface.   The drainage of a sloping field of uniform deep soil is a simple matter, the depth and distance apart of the tiles depending chiefly on the texture of the soil.   Particulars of this nature will be found in any treatise on agriculture and need not be dealt with here, since they are scarcely geological.

Difficult problems in drainage sometimes arise when the conditions all over the area dealt with are not uniform. For example, in low-lying ground it is sometimes observed that some patches are wet and in need of drainage, while other parts of the same field may be sufficiently dry. This generally arises from the occurrence below the soil of an uneven layer of impervious material, such as boulder-clay. The worst instances occur where the land includes the sites of ponds or meres, now dried up and filled with alluvium. The former existence of a pond here implies defective drainage, a condition unlikely to be ameliorated by natural processes. Similar effects may be brought about by local development of *pans* (see p. 140). In such places every drainage scheme has to be contrived to suit the local conditions and each dried-up pond must generally be dealt with separately. The former existence of these can often be demonstrated by the occurrence on their sites of masses of fresh-water shells; of such a nature are the well-known deposits of white shell-marl that mark the sites of the former "meres" of the Fenland. Occasionally it so happens that such areas can be drained by the simple process of digging a hole, called a dumb-well, deep enough to penetrate to a pervious stratum below. This process can also sometimes be employed successfully to get rid of small surface streams on heavy clay land; if the clay is not thick it may be possible to construct an artificial "swallow hole" down which the stream can flow. It is necessary however to be very careful not to divert polluted water into a water supply system by this method.

Ponds. A pond is essentially a hollow in the ground with an impervious bottom and sides, intended for storage of water. The simplest kind of pond is one excavated in a bed of some impervious material, such as clay; here no further preparation is required beyond the making of the excavation—the pond will hold water naturally, and will maintain itself so long as the supply of water from rain or from springs is at least equal to the natural evaporation and the artificial consumption. But if the excavation is to be made in porous material some means must be adopted to render the pond watertight. This is usually effected by the process known as puddling with clay, or by making some sort of stone pavement. For puddling any kind of stiff heavy clay will be found effective; if it is free from hard lumps or stones, so much the better; stony boulder-clay should if possible be avoided, since the stones are apt to work out and leave holes in the floor. If the pond is to be floored with stones or flags these should be thick and heavy in order to avoid displacement from the trampling of live stock. The joints between them can usually be made sufficiently water-tight by ramming with clay. Cemented ponds are as a rule only constructed at the homestead or near buildings, since they are expensive to make.

The water supply of ponds may be derived from various sources; if the excavation is made deep enough to intersect the water table, the supply will be dependent on the level of the latter. In most places this is fairly constant, though usually rising somewhat in winter and sinking in summer. If the pond is not deep enough it will dry up in summer. Where it is not possible to excavate down to the permanent water-level, the supply from rainfall may be sufficient to last the whole year, especially if surface water from surrounding slopes runs in freely. Again, it is very often possible to utilize some small surface stream, however insignificant, by means of an open ditch or pipes, or even the drainage water from some high-lying tract may be led into the pond. The construction of reservoirs on any considerable scale does not usually come within the scope of agricultural practice, though it is obvious that it is impossible to draw any hard and fast line of distinction between a pond and a reservoir. If it is desired to construct a dam across a stream, care must be taken to see that the foundation is suitable. A dam of any size can only be constructed with success on a foundation of some hard rock, though smaller dams may be founded on a stiff, tough clay; sand, gravel or alluvial deposits are all unsatisfactory, since the water will always work round or under the dam.

Dew-ponds. The water supply of upland regions lying on the Chalk and other limestone formations presents special difficulties. From remote times there have existed on the

Chalk downs of England ponds of a special type, known as "dew-ponds," which maintain a never-failing supply of water under the most unpromising conditions, and there has been much controversy as to the real source of the water supply, which is obviously not due to springs, since such are non-existent on the tops of Chalk hills. This problem was discussed in 1776 by Gilbert White of Selborne[1], who drew attention to the prevalence of heavy mists on these uplands, even in the height of summer, while the popular name, "dew-pond," testifies to the general belief that the water is derived directly from the atmosphere by a process similar to the deposition of dew on plants. Of late years it has been shown that the rainfall on the higher parts of the Downs is really much greater than on the lower ground in the same neighbourhood, and this may be of more importance in keeping up the supply than is commonly supposed.

Dew-ponds are usually constructed as follows: an excavation is made from 30 to 60 feet in diameter, 4 or 5 feet deep in the middle, with sloping sides. This is first lined with a layer of straw or brushwood and then a layer of puddled clay is placed on the top of this. It is generally supposed that the straw acts as an insulator, preventing the access of the heat of the ground to the layer of clay and to the water, which is cooled by radiation and thus condenses the dew. However in Wiltshire the straw is often put on the top of the clay, instead of below, and the ponds thus constructed seem to work equally well. It is evident therefore that something still remains to be explained. It is also stated that newly-made dew-ponds require to be artificially filled at first, and cannot be supplied by condensation of moisture on the dry surface of the clay or straw only, or even from the rainfall. Hence we must suppose that condensation on the cold water-surface is at any rate an important factor, if not the only one concerned.

Rainfall in the British Isles. Since the question of water supply is so intimately connected with the amount and distribution of rainfall, it may be well to consider the latter a little more fully. Within the limits of this small country the rainfall

[1] *The Natural History of Selborne,* Letter XXIX, second series, 1789

shows remarkable variations in amount, and this variability is directly attributable to the physical structure and to the geographical conditions. The British Isles lie in the temperate zone, in the region of prevailing south-westerly winds. This position is in itself favourable to abundant rain, since the air currents which have picked up much water vapour from the tropical Atlantic are here entering a cooler region, and thus becoming able to carry less water vapour, that is to say, they are approaching the point of saturation, when the water is condensed and becomes visible. Hence under any circumstances we should expect a copious precipitation in this region. But from special causes, this precipitation is most conspicuous on the western side of the country; most of the high and mountainous land of Britain lies to the west. The moist winds from the Atlantic blow against the high ground and the air is driven upwards, thus being cooled and becoming saturated[1]. Thus a heavy rainfall prevails on the mountains, and chiefly on the seaward side. When the air descends on the other side it becomes warmer again, and also actually contains less moisture, since it has parted with so much as rain. The west winds therefore on the eastern slopes are dry winds, tending to absorb moisture instead of depositing it. Hence they may also be described as "drying" winds. It follows therefore, from geographical considerations, that the climate of eastern England and Scotland must be much drier than the climate of the western parts of the same countries.

Along the eastern coast of both England and Scotland, at certain seasons of the year, easterly and south-easterly winds bring a good deal of rain, but when the east winds have crossed the main watershed and descend to the west they are dry and, being warmed by the descent, they bring the finest weather. This is especially the case in spring and early summer. On the east coast again a good deal of moisture is deposited by the sea fogs, when, although it is not actually raining, the air is saturated with moisture, which is condensed on cool surfaces, especially on plants. This phenomenon, which is known in

---

[1] The amount of invisible water vapour that air can contain is directly proportional to the temperature.

Scotland as a "haar," contributes materially to the growth of heavy crops in the east of Scotland generally and especially in the highly fertile districts of the Lothians, the Carse of Gowrie and Aberdeenshire. The cool moist atmosphere thus produced is particularly favourable to the growth of turnips and other root crops. In east Yorkshire also the root crops are as a rule heavier on the higher ground than on the lowlands, owing to the prevalence of sea fogs in such localities during the earlier part of the summer, when the newly germinated seedling turnips on the low ground are liable to be destroyed by drought and the turnip-beetle.

When we come to deal with still smaller areas the same considerations hold, though in a lesser degree, and in general terms it may be stated that hilly country has a higher rainfall than plains, and the western slopes of the hills have more rain than the eastern slopes. From the multiplication of observing stations in recent years it is becoming clear that these differences are considerably greater than was formerly believed to be the case, especially with regard to the influence of elevation on rainfall. A difference of a very few hundred feet in level may make a vast difference in the records of a rain-gauge. It is a matter of common observation how frequently thunder-storms follow even slight ridges, and the same is true in a less degree of the general annual rainfall. In a highly civilized country like Britain however another factor has to be taken into account, namely, the prevalence in certain areas of a smoky atmosphere. It is a well-known fact that in some large manufacturing towns the rainfall is disproportionately great, and this influence extends for some distance into the country districts round them.

So far as the actual distribution and amount of rainfall is concerned, we may say that Ireland is very wet, except in the south-east corner; the western half of Scotland is also very wet, as also are the higher parts of the eastern half. If we draw a line from the mouth of the Tyne to Bristol and then to Exeter, the remaining country is divided into two parts, wet on the west, dry on the east. This line is of course not perfectly definite, but approximate only, and there is a long dry area projecting into the Cheshire plain, mainly owing to the low general elevation of this tract. In general terms it may be stated that to the west of this line the average annual rainfall is more than 30 inches, to the east of it less than 30 inches. But in many parts of the west of England the rainfall is far higher than this; even in cultivated and low-lying districts it may reach 60 inches, while among the mountains the figure may reach twice this amount. The highest figure for many years anywhere recorded by trustworthy instruments is at Seathwaite in Cumberland, where the annual average is about 130 inches and in wet years over 200 inches have been measured. On the other hand, in the low-lying districts bordering the east coast, especially in Lincolnshire and Essex, the climate is very dry. At some places on the Thames estuary the annual average is as low as 18 or 19 inches, while round the Wash it is not much more. Over eastern England generally, south of the Tyne, the rainfall may be taken at about 25 inches. This would appear at first sight to be scarcely sufficient for successful agriculture, and in fact this district in certain years, such as 1911, does suffer from prolonged droughts. But in most years it is found that a considerable proportion of the rain falls in the summer months; in fact over a large area in Suffolk, Essex and Cambridgeshire, July is the wettest month of the whole year; this is probably to be attributed to the prevalence of heavy thunderstorms at that time. These are often accompanied by hail of such large size as to be seriously destructive to crops. Hailstones as large as a pigeon's egg, or even a small hen's egg, are not uncommon.

It is evident that variations of such extent, amounting to three or four hundred per cent. of the rainfall, must have an important influence on agriculture. A heavy rainfall favours the growth of most crops, but on the other hand it renders harvesting difficult and precarious. The wet districts are on the whole later than the dry districts, though the crops are often heavier. This applies more particularly to grass, green crops and roots. Wheat on the other hand prefers a hot dry climate, if the soil is not too light, and barley is certainly of better quality if the air is not too moist. But the most important influence of rainfall, although this is a fact not yet generally recognized, is on the soil itself, and especially on the stores of available plant food.   The researches of foreign investigators, especially in Russia, have shown how far-reaching is the influence of "washing out" or "leaching out" of the soluble constituents of the soil in moist climates.   This constitutes for example the special characteristic of the Podzol type of soil, which is so widely spread in Russia (see Chapter VI), and according to Ramann the soils in the western part of the British Isles approximate to this type, while in the east they belong to the "brown-soil" type[1].   Although the differences here alluded to are not very well marked in this country they probably exist.   The chief lesson to be learnt from these considerations is the desirability in a moist climate of keeping the soil well occupied by crops, avoiding fallows as much as possible, since when the soil is bare there must necessarily be a great loss of soluble material, especially nitrogen and lime.   On the other hand in a dry climate a bare summer fallow may actually lead to an increase of plant food by the rise of solutions from below.   In extreme cases such rising ground-water may bring up deleterious salts, as in the alkali soils of western America and other arid regions, but in Britain this does not happen.

[1] Ramann, *Bodenkunde*, 3rd edition, Berlin, 1911, p. 582.

CHAPTER VIII

GEOLOGICAL MAPS AND SECTIONS

In agricultural geology, no less than in other branches of the subject, the study of maps is of first rate importance. From an ordinary topographical map, without geological lines, it is possible to obtain much information of use to the farmer with regard to position, elevation and aspect of the land, drainage and water supply, accessibility by road, rail or river, distance from market towns and centres of population and many other points that may be of interest in view of a possible tenancy or ownership in a new district. Before inspecting a farm a good map should always be consulted, as it may be at once apparent that some insuperable drawback exists, or that the situation offers special advantages. All this is not geological and does not need much expert knowledge. But the study of geological maps with a view to forming an opinion as to the possible quality of the soil is a different matter, and requires some preliminary acquaintance with the subject.

Geological maps may be actually misleading as to the character of the surface soil, unless certain important principles are carefully borne in mind. The first point is in regard to variations in the lithological characters of rocks of the same age when followed from place to place.

In nearly all maps the rocks of a given age are indicated everywhere by the same colour or sign, regardless of change in their characters. Again, some maps show the superficial deposits as defined in Chapter v, while others ignore them altogether; another class of maps shows some of them and not others; this is perhaps the most misleading of all, though unfortunately very common. It is hardly necessary to insist on the obvious fact that the farmer is most concerned with the uppermost layer of the earth's crust and that what lies beyond the reach of the roots of crops is of secondary and indirect interest only. Then again most geological maps are on too small a scale to indicate the variations that are likely to occur within the limits of a single farm, at any rate in this country, where really large farms are exceptional. In spite of all these drawbacks, however, the study of geological maps when properly understood, is of great use to the landowner and farmer and may afford valuable information on certain matters of great practical importance, such as drainage, water supply, the occurrence of building-stone, road-metal, limestone, chalk, marl, gravel and so forth, as well as with regard to the probable nature of the soil.

Topographical maps. Of these there are almost innumerable varieties, on all scales. In some of them no attempt is made to indicate the relief of the surface, little being shown besides coast lines, rivers, railways, roads, towns, villages and political and municipal boundaries of various kinds. To this category belong most of the maps in atlases. In some of these the relief is partly indicated by the method of *hill-shading*, a term which explains itself. Most of these are on a small scale, indicating geographical position and little more. Of much greater use are the maps which show the relief of the surface by the use of contour-lines, and a special variety of contour maps, known as the *layer system*. A contour-line is an imaginary line passing through all points on the surface that are at equal heights above sea-level, usually, in this country, some multiple of one hundred feet. In the layer maps the spaces enclosed between two successive contour-lines are coloured with a tint corresponding to a definite range of height. For example in the excellent maps published by Messrs Bartholomew & Co. of Edinburgh, on the scale of 2 miles to 1 inch all the land up to 400 feet is coloured in shades of green, the deepest tint at the lowest levels. Above 400 feet brown tints are employed, increasing in depth upwards. By this means the relief of the land and the distribution of hill and valley are well brought out, and even on this small scale it is easy to make out the height and aspect of any elevated or inclined land, and even the steepness or otherwise of the slopes, by taking into account the narrowness or breadth of the colour bands. It must however be remembered that the intervals between successive contourlines are not constant, but usually increase at higher elevations.

For most purposes the best maps are those published officially by the Ordnance Survey of the United Kingdom. Of these the most generally useful are those on the scales of 1 inch to the mile, 6 inches to the mile and 25·344 inches to the mile. The 1-inch map, which can be obtained with contourlines or with hill-shading, is useful for a general view of a large district, showing roads, railways, towns and villages. The 6-inch map may be used for plans of large estates, but is too small for details. This map also shows contour-lines, at intervals of 100 feet up to 1000 feet and above this at wider intervals. The 50 foot contour is also shown. The 6-inch maps of Yorkshire and Lancashire are contoured at every 25 feet, instead of 100 feet, and these show small variations of the surface very accurately. For plans of single farms the third variety, commonly called the 25-inch map, is the most useful. On this the acreage of every field is indicated, but there are no contour-lines. On these maps individual fields are large enough to afford space for notes as to the succession of crops or any other information likely to be of value, and every field and enclosure is numbered. This map is also excellent for plotting the results of detailed soil-surveys. There are also maps on a larger scale than this, but these are only available for towns.

Geological maps. There are in existence many small-scale geological maps of the whole of the British Isles, of England and Wales, of Scotland, of Ireland, and of most other civilized countries. The majority of these are, however, apart from their small size, too generalized to be of much service to agriculturists, especially as very few of them take any cognizance of the superficial deposits. They serve only to give the most general idea of the geological formations that underlie any district, and afford little information as to soils. Small maps of similar character for special areas are also to be found in many county histories, county geographies and similar publications[1].

The most generally useful geological maps of the United Kingdom are those published officially by the Geological Surveys of the United Kingdom and of Ireland. Maps of the whole of the British Isles (except some small portions of the Highlands of Scotland, still incomplete) are published on the scale of 1 inch to the mile, and of these there are two series: the "solid" maps show only the rock-formations, ignoring the superficial deposits, except certain large areas of alluvium and peat; the other series, the "drift" maps, show the superficial deposits, paying special attention to the accumulations of glacial origin, and to the gravels of post-glacial date. In spite of some inconsistencies of treatment, especially among the earlier published maps, these are most useful for purposes of soil study, although the scale is small. Maps of a considerable part of southern England and some other special regions are now issued in a new and revised colour-printed edition. These are cheap and excellent.

Geological maps on the scale of 6 inches to the mile have also been published for certain districts, chiefly those where mining is important. Manuscript 6-inch maps of the whole country have been prepared and are preserved in the Museum of Practical Geology, Jermyn Street, London, where they can be consulted on application.

A so-called "Index Map" of England and Wales is also published by the Geological Survey, on the scale of 4 miles to 1 inch. This is now undergoing revision, and some sheets are now published in a "drift" edition. These show a wonderful amount of detail considering the small scale[2].

For different kinds of maps various methods of projection are employed, according to circumstances, varying with the size of the area depicted. All large scale maps are projections

[1] Special mention may be made of the excellent small geological maps given in the Cambridge County Geographies, published by the Cambridge University Press.

[2] A complete catalogue of the publications of the Ordnance Survey and of the Geological Survey can be obtained from Messrs Edward Stanford, Long Acre, London, W.C.

on a horizontal plane of a portion of the earth's surface as it would be seen from an infinite distance. Hence if it is required to indicate the relief of the surface the method of contour-lines must be employed. Let us consider for the sake of simplicity a region of moderate relief and elevation, for example, some part of England less than 1000 feet above sea-level. The published 6-inch Ordnance maps of such a region show contour-lines at intervals of 100 feet. Let it be assumed also that the region is free from superficial deposits, such as alluvium, gravel or peat, and consists wholly of stratified sedimentary rocks.

The first and simplest case is where the strata are horizontal; here it is evident that the outcrops will be parallel to the contour-lines, though the dividing lines between the different strata do not necessarily coincide with contour-lines; in fact they commonly do not. When the outcrops in such a map are coloured with appropriate tints for each stratigraphical division the general effect resembles the "layer maps" already mentioned. A close approximation to this type is to be seen in some of the published maps representing parts of the Carboniferous area of west Yorkshire, or parts of the Cretaceous outcrops in the south of Cambridgeshire. The width of the outcrop of a particular stratum as projected on the map is due to a combination of two factors, namely,

(1) the actual vertical thickness of the bed,

(2) the angle of slope of the ground.

If the surface of the ground is uniformly horizontal, only the uppermost stratum can appear, and the width of its outcrop is indefinite, being determined only by the area of the plain. On the other hand, in a vertical cliff, all strata below the highest will be invisible on the map, or as it may be otherwise expressed the width of their outcrop is zero. Between these two limiting values of zero and infinity the outcrop may have any width whatever. In practice, since most ground is undulating and most strata not very thick, the width of the outcrop is usually confined within fairly narrow limits. Since the three quantities, width of outcrop, slope of ground and thickness of stratum are connected by a definite mathematical relationship, knowing any two of them, we can calculate the third.

The problem may present itself in the following practical form: In the figure let *ABC* represent a section through a hill composed of horizontal strata of varying character; the surface from *A* to *B* is covered by soil and rainwash, so that no solid rock is visible; at *C* a well has been sunk and the thickness of the beds noted; it is required to find the width of the outcrop of the shaded bed on the slope *AB*. From the construction it is clear that the width of outcrop required when plotted on the map is given by *D'E'*, whereas when measured on the ground from *D* to *E* it will be slightly greater, in a ratio determined by the angle of inclination of the hill; the steeper the hill, the greater this difference will be[1]. This method may give valuable information as to the best place to open a quarry, and in many other cases. It is evidently equally applicable with modifications to inclined strata.

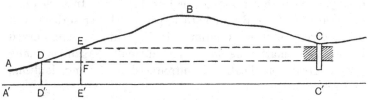

Fig. 41. *ABC* is the outline of the hill, seen in section, *A'C'* a horizontal line drawn at any convenient distance below the surface, *C* is a well, *D* and *E* the limits of the outcrop of the stratum, *D'E'* is the width of the outcrop as plotted on the map.

A similar plan can be employed in undulating country, composed of a succession of hills and valleys; the outcrop of a bed can be continued from one hillside to another, across an intervening valley, and when the strata are inclined such a section, if correctly drawn, will show whether a particular stratum should occur on the slopes of a neighbouring hill, or whether the dip has carried it either below ground-level or up into the air above the summit of the hill. For this purpose, as indeed in all other instances here mentioned, it is of course obvious that the section must be drawn to true scale, and as accurately as possible. The dips also must be measured with

[1] If the inclination of the slope *DE* to the horizontal plane be $\theta°$, then since $D'E' = DF$, therefore $D'E' = DE \cos \theta$; similarly $EF = DE \sin \theta$.

care and plotted with a protractor. The instrument commonly used for measuring the dip of strata is called a clinometer: there are many forms on the market, the simpler ones being the best. The essential feature is a straight edge fixed to a graduated circle, marked in degrees, with a movable plumb-line or pendulum. The straight edge is laid on the inclined rock-face, down its steepest slope, the angle made with the vertical by the plumb-line being read off directly from the circle. The clinometer is often combined with a compass for determining (by separate observations) the direction as well as the amount of the dip.

Hitherto it has been assumed that the thickness of the various strata, as well as their dips, if inclined, remains uniform, but of course this is not always the case. Many rock-beds are lenticular in form and of limited extent, thinning out and disappearing within a greater or less distance. In such cases observations of the dip of both upper and lower surfaces will often give useful information. If the surfaces are curved however the matter becomes more difficult. Again in many regions great complication is introduced by folding, faulting, and other dislocations.

In horizontal strata the only disturbances likely to occur are due to faulting. This is a complex subject, but one or two simple cases may be considered briefly. The general principles of faulting have been discussed in Chapter I, where the terms in use are also defined. The easiest case is where the fault is vertical; on sloping ground the effect of this will be to bring the outcrop to a sudden end; on level ground also this is the only thing that can happen. On a slope however the direction and amount of the throw of the fault may be such as to cause a repetition of the outcrop. The two cases are illustrated in Fig. 42. From these diagrams it is clear that when the upthrow of the fault is on the left or down-hill side, there can be no further outcrop unless the ground rises again in this direction. When the downthrow of the fault is on this side the outcrop of the bed may be repeated, or it may drop so far below ground-level as to disappear altogether. However if it does this the bed might be reached by boring. In nature it is by far the more

common to find the downthrow on the down-hill side, the
converse case being rather rare.   It is often possible, by studying
the strata above and below the one employed as an indicator,
to ascertain what has happened, since some of them are bound
to come to the surface somewhere on both sides of the fault,
although as a little consideration will show, a particular stratum
may be "faulted out" altogether, that is to say, it may never

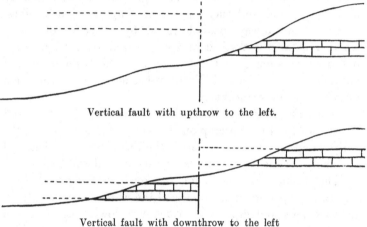

Vertical fault with upthrow to the left.

Vertical fault with downthrow to the left
Fig 42.

intersect the surface at all.   When dealing with problems
connected with faulting the methods briefly outlined above can
be applied successfully to the solving of the simpler problems.
The effects of faulting of various kinds in repeating and
suppressing the outcrops of inclined strata are dealt with in
Chapter I (see p. 23 and Figs. 12 and 15).

When dealing with inclined strata, in addition to the slope
of the ground and the thickness of the beds, we have to take
into account also the direction and amount of the dip.   In
inclined strata the thickness is not measured vertically, but it
is the least distance between the upper and lower surfaces of
the bed, perpendicular to the bedding planes whatever the
inclination (dip) may be.   If the beds are vertical therefore the
thickness is measured horizontally.   As a matter of convenience
vertical strata may be considered first.   Here outcrop and

strike[1] coincide and the width of the outcrop, as plotted on the map, can never be less or greater than the true thickness of the bed.   The outcrops of vertical strata will of course undulate in the vertical plane when followed over hills and valleys, but when correctly projected on a map they must appear as straight parallel lines, so long as the thicknesses remain uniform.   Exactly similar statements apply to fault-planes; a straight vertical fault must appear on a map as a straight line; many faults are however curved, and such appear on maps as sinuous lines, hence it is not always possible to tell by inspection of a map whether a fault is vertical or not; a perfectly straight fault indicated on sloping surfaces must be vertical, but a curved outcrop of a fault may indicate either actual curvature or a departure from verticality.

When the strata are inclined at any angle between 0° and 90° the width of the outcrop on the map varies indefinitely, the three independent variables being the thickness of the bed and the angles of inclination of the surface and of the bed.

The simplest case is where the ground is horizontal, since one variable is here eliminated, and the width depends only on the thickness and dip, being wider with lower angles of dip. Fig. 43 shows the relations between thickness, dip and width of outcrop in an inclined stratum.

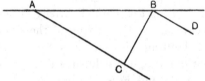

Fig. 43. BD = upper surface of bed, AC = lower surface, BC = thickness of bed, AB = width of outcrop, BAC = angle of dip.

Again when the dip and thickness are uniform it is easy to show that the width of the outcrop as seen on the map depends on the slope of the ground, being widest on a level plain and vanishing on a vertical face.   If the bed dips the same way as the surface of the ground, the width of the outcrop is indefinite.

[1] For definitions of dip and strike see p. 17.

When dealing with inclined strata outcropping on inclined ground there are thus three variable factors to be taken into account, namely, dip, thickness, and slope of the ground. Since however these are obviously connected by simple mathematical relations, if any two are known it is easy to find the third by calculation or by graphic methods; the latter are generally employed. The question evidently becomes more complex when the slope of the ground is variable, and especially when the ground slopes the same way as the bed. Each such instance must generally be dealt with independently and no general rule can be given. The only method that gives a reliable result is to make a profile section of the ground to true scale, from the contour-lines, and then to insert the geological structure, with the correct dip and thickness; the section will then show where any particular stratum may be looked for, if its outcrop is covered by drift, rainwash or any other superficial accumulation. This method will often also give useful indications of the probable thickness of such superficial deposits at any given point. For instance, if there is reason to believe that an old valley has been partly filled up by later deposits, thus giving steep sides and a flat floor, it may be possible by continuing downwards the slope of the steep upper part, to obtain a rough estimate of the probable depth from the surface of the old sides of the valley. Such a section, to be of any use, must evidently be drawn to true scale.

The next point to be considered is the form of the outcrops of inclined strata on undulating ground. It has already been shown that vertical strata have straight outcrops, while the outcrops of horizontal strata are parallel to the contour-lines; hence for any intermediate inclination the outcrops must be sinuous lines, cutting the contour-lines. The direction of the sinuosities depends on the form of the surface. The relations are most easily seen by means of a solid model representing a river valley carved out of inclined strata. When the strata dip towards the head of the valley their outcrops will have a V-shaped form, the apex of the V pointing up the valley; when the dip is down stream, at an angle greater than the slope of the stream, the apex of the V will point downwards.

When the dip of the beds is down the valley at an angle less than the slope of the stream, the apex of the V will however point up stream. This case is less common than the other two. By taking account of these facts it is usually easy to determine by inspection the direction of dip of inclined strata.

When a geological map showing contours is drawn accurately to scale it is possible to determine by purely graphical methods the dip and thicknesses of the strata shown, to find the depth from the surface at which a given bed will be found at a particular place, to measure the throw of faults and to obtain much information of value to practical men. It is impossible however, owing to limitations of space, to enter here into a description of the methods to be employed.

CHAPTER IX

STRATIGRAPHICAL GEOLOGY. INTRODUCTION

For the agriculturist the most important application of geology is the study of the geographical distribution of the different types of rock and their relations to crops and stock. It is well known to every one that in different parts of any given country different kinds of farming prevail; in one area nearly all the land is under the plough, in another it is mainly permanent pasture, while in other areas again, especially high-lying and mountainous regions, the land is mostly unenclosed, being devoted to grazing, or even entirely unoccupied by domesticated animals, though in Britain this condition is rare. In determining the agricultural potentialities of a district geological structure and the characters of the rocks are of predominating influence, though other factors play an important part, such as climate, economic conditions, and even the mental and moral characteristics of the people. The effect of climate is undoubtedly very great and in the British Isles it is specially well marked, owing to the great differences in the temperature and rainfall on the eastern and western sides of the country. The wet climate of the western side is largely attributable to the superior elevation of the land, and this is due in the first place to the geological structure, most of the older and harder rocks occurring on that side and tending to form high ground. Thus it is clear that some of these factors are mutually interdependent, and it is often very difficult or even impossible to disentangle their effects. The influence of latitude has also to be taken into account, although it appears that in the temperate zone at any rate, this is comparatively small. For instance in central Norway, only a very few degrees from the arctic circle, luxuriant forests grow and farm crops ripen up to a much greater height above sea-level than in Scotland or Ireland, 8 or 10 degrees further south. In the warmer regions of the world, in the tropical and subtropical zones, the rainfall is the chief factor, and in no subject more than in agricultural geology do we realize the importance of the existence of climatic belts, as explained in Chapter II. However rich a soil may be, if dry it is barren. Hence the enormous importance of irrigation in Australia, South Africa, and western America, where there are often great stores of underground water waiting to be tapped, or streams running from mountain ranges through sandy wastes, and only needing to be properly distributed to yield surprising results in the way of heavy crops from soils of most unpromising appearance.

To sum up, the suitability of a region for farming depends not only on the character and distribution of the rocks, but also on latitude, climate and other factors of minor importance.

The object of the present and succeeding chapters is to study the characters and geographical distribution of the rocks and other deposits that compose the outer crust of the earth, forming the surface of the ground and giving rise by their decomposition and decay to the cultivated soil on which the farmer grows his crops and pastures his stock. Since this book is primarily intended for English readers most attention will be paid to the rock-formations as developed in the British Isles, but incidental reference will be made to foreign and colonial examples in illustration of special points.

The agricultural importance of any particular formation depends on a large number of independent and variable factors. Thus for example a thick succession of rocks in a vertical or highly inclined position may form a much less area of the surface of a country than a thin series spread out in horizontal or very slightly inclined layers; again a thin series, though highly inclined, may be repeated many times by folding. But of still more importance is the fact that the solid rocks of the crust are often concealed from view by a layer of transported material, drift or alluvium; in such a case it is the nature of the super-ficial layer that is of interest to the farmer, since this determines the nature of the soil. It is only the uppermost layer of all that forms the cultivated soil; however rich in plant food the underlying materials may be this is useless if covered up by a very few feet of unworkable or unproductive material. A striking example of this is afforded by the surface quartzites and surface limestones of South Africa and other dry regions, described on p. 117.

A consideration of such facts shows at once the critical importance of the careful investigation and mapping of the superficial deposits, and the comparative neglect of these has rendered many geological maps worse than useless to the farmer. A map showing only what the geologist calls the *solid* formations is merely misleading, since the strata shown on the map may actually be buried under tens or even hundreds of feet of superficial deposits of entirely different character. It is only within recent years that true soil-surveys, confined to the workable soil itself, have been carried out in certain limited areas. The most useful maps for agricultural purposes are the "Drift" series published by the Geological Survey, but even in these the scale is only 1 inch to the mile, and very different types of superficial deposit are often indicated by the same colour. A simple designation, such as boulder-clay, includes a great variety of clays and more or less sandy beds formed by glaciers and ice-sheets, while gravels and alluvium often differ a good deal in composition and agricultural character. It is only in the south and south-west of England that glacial deposits are wholly absent over large areas and here the "solid" maps indicate with precision the character of the soils.

After due allowance has been made for all the difficulties detailed above, the fact remains that a geological map on a sufficiently large scale does afford a general indication of the kind of soil that is to be expected in any district, and if along with it we study a topographical map, showing the relief of the surface and consequently what is commonly spoken of as the aspect of the farm, very useful information is obtained. This last consideration is an important one, and often over-looked. It is obvious that of two farms lying on similar rocks, but one sloping northwards and the other south, the one on the southerly slope is more likely to grow good crops and to rear healthy stock. This is for the most part a matter of temperature and climate, and not really geological, but in hilly and still more in mountainous regions it is of the greatest importance. Again exposure to or shelter from the prevailing winds has great effect on the character of the crops, especially near our eastern coasts. Hence it appears that from many points of view the study of maps is of great agricultural importance. The dependence of water supply on the underground structure of the rocks has been dealt with in an earlier chapter (see p. 166), and it was there clearly shown that an acquaintance with the succession of the stratified rocks is essential for a successful attempt at water-finding.

The rocks composing the accessible portions of the earth's crust have been divided by geologists into somewhat arbitrary groups, each of which is characterized by some special feature distinguishing it from all others. The stratified rocks are subdivided partly by structural features and lithological characters, but more satisfactorily by means of their included fossils. Unfortunately the oldest stratified rocks do not contain any fossils, and it is sometimes difficult or impossible to determine their true sequence and the mutual relations of detached portions. For this reason it has been found necessary to adopt an arbitrary base-line and from this to work both upwards and downwards. The oldest definitely established fossils are found in a certain set of rocks which have long been called the Cambrian system, and the base of this system is taken as the primary datum-line; all the rocks demonstrably older than this being called Pre-Cambrian. It is claimed that in America and elsewhere fossils have been found in rocks older than the Cambrian, but even if this be so, the general principle is not affected. The rocks both above and below the datum-line are divided into *systems* and these again into *series* and even smaller units, as will be hereafter described. The major divisions, the boundaries between the systems, are frequently determined by the more important unconformities, or breaks in the succession, while the minor subdivisions depend either on lithological or on palaeontological variations. Many

Table of the Stratigraphical Rock-systems.

Groups	Systems
Kainozoic or Tertiary	Neogene Palaeogene
Mesozoic or Secondary	Cretaceous Jurassic Triassic
Palaeozoic or Primary	Permian Carboniferous Devonian Silurian Ordovician Cambrian

Pre-Cambrian

The systems, as defined in the right-hand column, are divided into series and these again into stages; in very detailed work even smaller subdivisions are employed. The Pre-Cambrian group includes an enormous thickness of rocks, whose mutual relationships are still uncertain; several distinct systems are recognizable. As their relative order of succession is in some cases unknown they are here omitted for the sake of brevity.

By many writers the Tertiary group is cut up into a larger number of systems than here shown; it is better however to regard these as having the value of series only, since they are much thinner than the older systems.

of these however are of purely local value, since the nature of rocks of the same age varies much from place to place, according to conditions of formation.

The table on p. 197 shows the principal divisions and sub-divisions of the stratified rocks as adopted by the majority of British geologists. The classifications in use in other countries run on the same general lines, and in many instances the same names are employed. This branch of geology may be said to have originated in this country from the classical work of William Smith, who applied to the different formations names taken from the localities where he had studied them. Hence it results that the names of obscure English villages, such as Kimeridge in Dorset, or even local dialect words, such as Gault, have a world-wide currency and have been adopted into most civilized languages. In our colonies and in many foreign countries a mistaken patriotism has led to the introduction of many other names, often equivalent to the original English terms and of no more general application, so that the whole subject has been involved in great confusion. The English names, though often unsuitable, have at any rate the claim of priority, while the innumerable foreign innovations are in many cases entirely unjustified and unnecessary. The completeness of the geological record in the British Isles is indeed very remarkable; it has been found in only a very few instances that formations recognizable in other countries cannot be identified in Britain, though often with a somewhat different character.

It is frequently found that a formation of a given age, when traced from place to place for a considerable distance, undergoes a gradual change in character, corresponding to an actual difference of physical conditions at the time of its formation. The contemporaneity of stratified rocks in two distant regions can be established in two different ways; it may be possible to trace the bed continuously from one place to the other, though this is rarely practicable; or the equivalence of age may be established by examination of the contained fossils, a much more convenient and certain method. If the bed in question itself contains no fossils, it may be possible to establish its identity by noting its relation to other strata of known age, and this is a method of wide application.

The principle that a stratified rock varies from place to place in harmony with the local conditions is most important, and has led to the development of the idea of *facies*, one of the most fruitful conceptions in modern geology. The facies of a formation may be defined as the sum of its lithological and palaeontological characters at any given place. These characters, when carefully studied and correctly interpreted, indicate the physical conditions under which the bed was formed, and thus afford valuable information as to the climatic and geographical conditions of past times. This subject is not of much direct interest to the agriculturist; but the study of facies, or lateral variation of strata, as it may be called, has an indirect interest, in the following way. It is generally the custom in geological maps to indicate all the strata of the same age by the same colour, irrespective of variation in character, and it is consequently unsafe to assume that the same colour in two distant parts of a map necessarily indicates the same *kind* of rock, although it does indicate rock of the same *age*. Many instances of this principle will appear in the following chapters.

Even within the narrow limits of the British Isles many of the rock-formations show a wide variation of character, and therefore of agricultural value, when traced from place to place, and this initial difference of composition is often made still more notable from the point of view of soil-formation by the wide range of climatic conditions that here prevail. Some formations are of such a character that they may yield good soils in a favourable climate, and bad soils where the conditions are adverse. In other instances the inherent characteristics of the rock are dominant over climatic disabilities and the soil may be conspicuously fertile or infertile under very varying circumstances. A good instance is the Old Red Sandstone, which yields rich soils, growing heavy crops even in the inhospitable climate of the extreme north of Scotland, as well as in the warm and sheltered Severn valley. This is an extreme case, but other less striking instances will appear later.

CHAPTER X

THE PRECAMBRIAN AND LOWER PALAEOZOIC SYSTEMS

I. THE PRECAMBRIAN SYSTEMS

The Precambrian rocks of the British Isles are almost confined to the Highlands and Islands of Scotland and to the north and west of Ireland. In these two countries they form a large extent of land, but in England and Wales the areas over which these rocks come to the surface are very limited, and of little or no agricultural importance.

Within the limits of the British Isles three very well-marked and distinctive types of Precambrian rocks have been observed, as follows: (a) gneisses and schists, (b) sediments, (c) volcanic rocks. It is quite certain that the gneisses and schists are the oldest of these groups, but the relative ages of the sediments and volcanics are uncertain. If a line be drawn across Scotland from Stonehaven on the east coast to Helensburgh on the Clyde and continued to the neighbourhood of Galway in Ireland, this line may be taken in a general way to mark the southern boundary of the visible gneissose and schistose rocks, though even over considerable parts of this area, and especially in Ireland, they are covered by more recent formations. Similar rocks reappear over a small area in Anglesey and a still smaller patch exists in the Malvern Hills. The sedimentary type occurs as a narrow and often discontinuous strip down the west of Scotland from Cape Wrath to Islay, reappearing again in Shropshire, while the volcanic type is found in disconnected patches in Wales, Shropshire and Leicestershire.

The Highland gneisses and schists include a great variety of rock-types of differing origin, some originally igneous and some originally sedimentary, but all agreeing in the fact that they have undergone intense pressure-metamorphism, whereby they have been wholly recrystallized and new minerals formed, thus completely changing their primitive character. The oldest rocks of all, forming the islands of the Outer Hebrides and a narrow strip on the adjoining mainland, are mainly igneous in origin. These are called the Lewisian gneisses, after the island of Lewis. They consist for the most part of granites, diorites and other more basic rocks, intensely sheared and traversed by hundreds of dykes, the latter being largely converted into hornblende schists. Only near Loch Maree are there some traces of sediments, including crystalline marbles. The general character of the scenery may be gathered from the following description of the gneiss country as seen on the mainland, along the sea-board of Sutherland and Ross. "Throughout this belt of country bare rounded domes and ridges of rock, with intervening hollows, follow each other in endless succession, forming a singularly sterile tract, where the naked rock is but little concealed under superficial deposits, and where the surface is dotted over with innumerable lakes and tarns[1]." It is obvious that such country can have little or no agricultural value, and it is mainly deer-forest.

Resting unconformably on a very uneven surface of Lewisian gneiss is a vast thickness of true sedimentary rocks, known as the Torridonian or Torridon Sandstone group. This consists mainly of red felspathic sandstones and conglomerates, with occasional bands of shale, showing a strong superficial likeness to the Old Red Sandstone. Its Precambrian age is however conclusively proved by the fact that in its turn it is overlain unconformably by fossiliferous Cambrian strata. The Torridon sandstone area is in the main mountainous, and it forms some of the most remarkable peaks in the British Isles. Judging from its general character this formation would under favourable

[1] "The Geological Structure of the North-West Highlands of Scotland, *Mem. Geol. Survey*, Glasgow, 1907, p. 2.

conditions yield a rich soil, but the climate and situation are so adverse that this also is in the main under deer forests.

As previously mentioned a very similar rock reappears in Shropshire, forming the high-lying district known as the Longmynd. This reaches a height of 1700 feet and is some 12 miles long by 5 miles wide. It is mainly open unenclosed ground, covered with heather and devoted to sheep-farming, the chief breeds being the Welsh Mountain, Clun and Kerry. There are also droves of more or less wild ponies. Only in the valleys is there a little arable land with a poor, sour soil.

The greater part of Scotland, north of the line as previously defined (p. 200), is occupied by a series of rocks which may be conveniently known collectively as the Highland Schists; these undoubtedly include in places patches of Lewisian gneiss, brought up by faults, and they are also largely penetrated by great granite intrusions, as in Aberdeenshire and near Oban. But the greater part of the area is occupied by a series of rocks, originally sediments and now intensely metamorphosed, so that they have taken on the structure of typical crystalline schists. Besides felspathic gneisses and mica-schists they include also quartzites, slates, and crystalline limestones or marbles.

The more highly felspathic gneissose rocks of Sutherland and Ross are often spoken of as the Moine gneisses, while the more definitely sedimentary rocks of the central and southern Highlands and of the north of Ireland are called the Dalradian series. From recent investigations it appears that the structure of this region is of almost incredible complexity and the true relations of the rock-groups are not as yet understood. However it is clear that the rocks are generally speaking thrown into folds striking from N.E. to S.W. and some beds, especially a coarse conglomerate, have been traced across Scotland from sea to sea. The limestones and quartzites appear again in force in Ireland.

Since both the rocks themselves and the climate show very wide variations within this area the agricultural characters of this formation vary enormously. The western parts are as a rule distinctly less fertile than the eastern areas in the same latitude; this is due partly to superior elevation and partly to excessive rainfall. The greater part of the land is mountainous, being largely deer-forests, grouse-moors and sheep-runs, but in all the valleys and along the east coast is much fertile land, though much of this is certainly on glacial drift and other superficial deposits. In the glens and sheltered places artificial plantations are successful, though in the Highlands natural forest is almost non-existent, and contrary to a wide-spread opinion, deer-forests do not contain any trees. Especially in the west, peat bogs are very abundant. Although there is much successful arable farming, as for example in Aberdeenshire, nevertheless the staple industry of the Highlands is certainly sheep-grazing, and the rearing of the well-known Highland cattle, which are well suited to a rigorous climate and poor food.

In Ireland the Dalradian rocks occupy a large area, generally at a less average elevation than in Scotland. They are much covered by peat and other superficial deposits, but where exposed at the surface they sometimes form what in Ireland is considered fairly good land. Some parts however, as in western Donegal and in Connaught, are very infertile, the climate making any real improvement impossible.

The Precambrian volcanic rocks of Wales and of Shropshire cover so small an area that no description is necessary. Charnwood Forest in Leicestershire consists of similar rocks, while the Malvern gneisses also cover very little ground. The serpentine and other igneous rocks of the Lizard in Cornwall, which are probably Precambrian, form a barren wind-swept plateau, partly covered with heather and gorse and bearing many rare plants.

In many other parts of the world Precambrian rocks, both crystalline schists and sediments, cover enormous areas, and yield soils of all kinds, their character depending largely on climatic conditions. The Scandinavian peninsula is very like the Highlands of Scotland on a larger scale; among numerous examples special mention may be made of Canada, South Africa and India as countries where the Precambrian rocks play a very large part in the structure of the land.

Where conditions are favourable the soils yielded by them are often very fertile. Hence the comparative barrenness of these formations in Britain is clearly in the main a matter of climate and elevation rather than any peculiarity inherent in the rocks themselves. Since gneissose and schistose rocks often possess a composition very similar to that of granites and other igneous rocks, they must be rich in many of the elements of plant food, only requiring favourable weathering conditions to yield rich soils. In many tropical regions such conditions exist to a very large extent. Hence arises the great fertility of Brazil and parts of central Africa, among many other examples.

II. The Lower Palaeozoic Systems

The Lower Palaeozoic systems of the British Isles comprise an enormous thickness of rocks showing a general uniformity of character, though there is much variation in local detail. For the most part they form elevated regions in the northern and western part of the country, the three principal areas being Wales, the Lake District and the Southern Uplands of Scotland. Since the total thickness is so great, they are divided by geologists into three systems, the Cambrian, Ordovician and Silurian. Unfortunately there exists considerable confusion of nomenclature, since by many writers the Silurian system is held to include the Ordovician, under the name of Lower Silurian, while the Silurian in the modern sense is called Upper Silurian. It is unnecessary to enter into this controversy; it must suffice to say that the old arrangement is now obsolete, and the Ordovician system is recognized by nearly all modern authors.

In general terms it may be stated that in the main the Lower Palaeozoic rocks consist of slates, grits and conglomerates with only occasional limestone bands, though these are often of some local importance. In places also there is an extensive development of igneous rocks of Ordovician age, as in North Wales and the Lake District. These volcanic rocks form some of the wildest mountain scenery in the British Isles, and only a very small proportion of the whole series is suited for high-class farming.

A. The Cambrian System

The Cambrian rocks, as the name implies, are most extensively developed in Wales, forming several detached areas in that country. The largest of these are in Caernarvonshire and Merioneth.

The subdivisions of the Cambrian system adopted in North Wales are as follows:

> Tremadoc Slates,
> Lingula Flags,
> Menevian series,
> Harlech Grits and Llanberis Slates.

The Harlech Grits form a high, generally uncultivated, area in the west of Merioneth; the soils are generally unproductive and most of the land is in mountain pasture.   Of the same age are the well-known roofing slates of Llanberis, which give rise to an important quarrying industry at Bethesda and other places in Caernarvonshire.   The Menevian strata, which are thin and much softer, generally come at the bottoms of deep valleys, but the Lingula Flags form a wide extent of upland country, much covered by drift and peat.   The Tremadoc Slates are also unimportant, though yielding an inferior grade of slate.

The Cambrian rocks of South Wales, which are closely comparable with those described, form some small patches near the city of St Davids in Pembrokeshire; there are also small outcrops in Shropshire, in the Malvern Hills, and near Nuneaton, in Warwickshire.   The Hartshill quartzite at Nuneaton yields a valuable road-making material much used in the midland counties.   In the north-west of Scotland there is a long strip of Cambrian rocks, extending near to the western sea-board from the neighbourhood of Loch Eireboll to Skye.   This includes quartzites and limestones or dolomites, some of the latter being metamorphosed to marble.

The Cambrian rocks contain the earliest known fossils, the most important group in this system being the trilobites. *Olenellus*, *Paradoxides*, *Olenus*, *Asaphus* and *Angelina* are the most characteristic genera.   Brachiopods also occur, and the earliest graptolites are found in the Tremadoc division. However fossils are scarce and often difficult to find, owing to cleavage.

Since the Cambrian rocks cover only very small areas in remote and often mountainous regions it is unnecessary to give any detailed account of their lithological character. Where cultivated soils are found on Cambrian rocks the subsoil is usually found to be drift and other superficial deposits.

B. The Ordovician System

The strata of the Ordovician system cover large areas in Wales, the Lake District, and the south of Scotland. They form for the most part elevated ground, often indeed rising into mountains. This system differs from the Cambrian in that it includes in North Wales and the Lake District great thicknesses of volcanic rocks, both ashes and lavas, but the sediments are of much the same general character as the Cambrian, being mainly slates and grits, with only local calcareous developments of no great thickness.

The rocks of the Ordovician system are divided into four groups, as follows:

> Ashgillian series,
> Caradocian series,
> Llandeilo series,
> Arenig or Skiddavian series.

The Caradocian and Ashgillian together are equivalent to the older designation of *Bala series*; the latter name has now been generally abandoned, owing to a want of precision in definition. The typical development of the system is seen in Merioneth and Caernarvon, though the rocks of Shropshire and the Lake District are perhaps equally important geologically, if not agriculturally.

In North Wales the Arenig rocks rest with a slight unconformity on the Cambrian; at the base is a conglomerate, with shales or slates above. In the Arenig mountains and near Dolgelly is a great thickness of lavas and ashes, mostly forming very wild uncultivated ground. The Llandeilo series is also

in the main slates, with some igneous rocks, but the two upper
subdivisions show more variety, being in some places an
alternation of slates and limestones, in others being represented
by the great lavas and ashes of the Snowdon district. Snowdon
itself is mainly composed of lava flows.

The Ordovician rocks of Pembrokeshire are closely comparable with those of North Wales, though igneous rocks are
less conspicuous. In most geological maps almost the whole
of central Wales is indicated as being composed of Ordovician
rocks, but in reality a good deal of this area is occupied by
Silurian strata, as shown by the recent detailed study of some
portions. However the two systems are here very similar, both
lithologically and agriculturally. Ordovician rocks also occur
to a considerable extent in Shropshire, both east and west of
the Longmynd. Here they form rather poor soils, mostly
heavy and wet, those on the east being better than those on
the west.

The Ordovician rocks of the Lake District may also be
divided into four groups, correlated with those of Wales; they
are as follows:

> The Ashgill Shale,
> The Coniston Limestone,
> The Borrowdale Volcanic series,
> The Skiddaw Slates.

The Skiddaw Slates include a great thickness of rather soft
slates, generally of poor quality, with some grits. However it
is known that part of the series is Cambrian, there being here
no visible break in the succession. The Skiddaw Slates form
all the northern part of the Lake District, including the Skiddaw-
Saddleback mountain mass and they extend far to the west,
even beyond Ennerdale. The higher parts are almost entirely
open sheep ground; in the valleys there is much drift and peat,
and the soil is on the whole poor. In the north the lower ground
is mainly covered with boulder-clay and the soils are very stony.
Besides the slate quarries there are lead mines, now largely
abandoned, and the iron-mining district of west Cumberland
extends into the Skiddaw Slate area east of Cleator Moor.

The Borrowdale Volcanic series, which is computed to have a total thickness of some 30,000 feet, forms the main part of the Lake District mountains, including the Scafell group and the Helvellyn range   Most of the country is exceedingly steep and rocky; only in a few valleys is there any cultivation, mostly on drift and alluvium, and the highest summits are bare rock. Sheep-farming is almost the only kind of agriculture. The chief economic products are the famous green slates (cleaved volcanic ashes) and the graphite of Borrowdale, used for lead-pencils.

The Coniston Limestone and the Ashgill Shales are both thin and their outcrop, though traceable for many miles from Shap towards the south-west, is too narrow to be of any importance.   There is also a narrow strip of Ordovician rocks in the Eden valley, along the foot of the Cross Fell range; this is largely open sheep ground.

The Ordovician system of the Lake District is the home of the Herdwick breed of sheep; apart from sheep-farming agriculture is not in a flourishing condition, largely on account of the excessive rainfall, and the mountainous nature of the ground.

Ordovician rocks also occur to a considerable extent in the Southern Uplands of Scotland, but they are there inextricably mixed up with the Silurian, both formations showing similar lithological characters in the same areas, though varying much from place to place; hence consideration of this region may well be deferred till the Silurian system is dealt with.

The Ordovician rocks are often fossiliferous, except of course where they are of volcanic origin.   The fauna is fairly large, trilobites being still abundant, though the graptolites are perhaps even more important.   By means of the latter group it has been found possible to divide the whole system into a series of minor subdivisions called *zones*, traceable in all the different areas, the same forms occurring in the same succession also in Scandinavia, North America and other parts of the world. The most important Ordovician genera are *Dichograptus, Tetragraptus, Phyllograptus, Didymograptus, Dicellograptus, Dicranograptus, Diplograptus* and *Climacograptus*.   Among the trilobites the most characteristic are *Asaphus*, *Ampyx*, *Ogygia*, *Phacops* and *Trinucleus*. Among the brachiopods are *Orthis*, *Strophomena* and *Leptaena*. Corals and the very remarkable cystideans are common in the limestones.

C. The Silurian System

As previously explained the term Silurian is here used in the modern restricted sense, equivalent to the Gothlandian of some recent continental writers. This includes only the Upper Silurian of the earlier British authors. In most parts of the British Isles, where Silurian rocks are seen at the surface, they present a facies very similar to that of the preceding Ordovician system, namely, a great thickness of shales and grits of very monotonous character; it is only in Shropshire and the Welsh borderland that limestones are developed to any extent. It is an unfortunate fact that in Shropshire, usually regarded as the type area, the development of the system is abnormal, with thick beds of limestone. Everywhere else the gritty and shaly facies is dominant. The names applied to the principal subdivisions are largely taken from the Shropshire district, consequently in most parts of the country they are entirely inapplicable.

The Silurian rocks crop out at the surface over considerable areas in Shropshire and Herefordshire, and form a large part of Wales. As already mentioned, they have not yet been properly demarcated from the Ordovician in central Wales. They also occupy large areas in the southern part of the Lake District and the adjoining Howgill Fells, and in the Southern Uplands of Scotland. In Ireland also there are many scattered patches.

Two distinct classifications of the strata are in general use, as shown side by side in the following table:

Downton and Ledbury beds ⎱ Upper Ludlow ... ...⎰	Downtonian series.
Lower Ludlow ... ...⎱ Wenlock series ... ...⎰	Salopian series.
Tarannon Slates ... ...⎱ Llandovery beds ... ...⎰	Valentian series.

The classical district for the Silurian succession is in Shropshire, in the country between Much Wenlock and Ludlow. Here the system rests unconformably on the Ordovician, the Lower Llandovery being absent. The Upper Llandovery is represented by the May Hill Sandstone, a calcareous sandstone with many fossils, especially *Stricklandinia* and *Pentamerus oblongus*. The Tarannon series, if present at all, is unimportant. At the base of the Wenlock series in some places comes the Woolhope limestone, which is inconstant, occurring here and there in lenticular masses. The greater part of the Wenlock series consists of a thick mass of black shales with graptolites of the *Monograptus priodon* group. At the top comes the well-known Wenlock Limestone, economically the most important bed of the whole system. This is largely quarried for lime-burning and for use in the iron-furnaces of the Black Country. It is a grey limestone, about 100 feet thick and extraordinarily fossiliferous, showing many of the characters of a coral-reef. Corals, crinoids, brachiopods and trilobites are very abundant and well preserved. Some important forms are *Favosites, Heliolites, Halysites, Acervularia, Pentamerus galeatus, Atrypa reticularis, Calymene Blumenbachii, Phacops caudatus, Encrinurus punctatus*.

The Lower and Upper Ludlow beds are lithologically very similar; the Lower Ludlow contains graptolites of the *Monograptus colonus* type, but this group became extinct at the top of this subdivision. The dividing line between the Lower and Upper subdivisions is formed by the Aymestry Limestone, characterized by *Pentamerus Knightii*. The Upper Ludlow consists of grey flaggy shales with *Chonetes striatella* and a so-called "bone bed," full of bones and spines of fish and other marine animals. The Downton beds are sandstones, gradually becoming more and more like the overlying Old Red Sandstone. The principal fossils are Eurypterids.

In North and Central Wales certain changes are seen in the lithological characters of the system, the Llandovery and especially the Tarannon series becoming much thicker; near Plynlimmon the latter attains about 3000 feet. The limestones disappear and the whole system becomes a very thick and

monotonous succession of shales and massive grits. In central
Wales numerous graptolite zones have been established, and
most of them recur also in the Lake District and Scotland.

In the Lake District the Valentian series is represented by
thin black shales with graptolites, the Stockdale shales, the Wenlock is flaggy, and the Ludlow series expands to an enormous
thickness, estimated at 14,000 feet, forming a great expanse of
poor high-lying ground, both in the southern part of the Lake
District and in the Howgill Fells, as far east as Sedbergh. In
Scotland, near Moffat, the Llandovery division alone consists of
black graptolitic shale, all the rest, so far as it is present, being
thick grits with occasional shale bands (Gala grits and Riccarton
series). These form most of the open high ground of the
Southern Uplands and are many thousands of feet thick.
Throughout this region the upper limit of the Silurian is somewhat uncertain, and only at Lesmahagow in Lanark is there any
representative of the red Downton sandstone. The Silurian
rocks of County Down are much like those of the Southern
Uplands; in the rest of Ireland they are mainly covered by
peat and drift.

Throughout the greater part of all these areas the Silurian
rocks form high ground, which is still largely unenclosed and
devoted to sheep-farming. In the valleys, where the soil is
sometimes fairly fertile and well cultivated, it lies for the most
part on drift and is not to be regarded as derived directly from
Silurian rocks, though these undoubtedly in most cases supplied
the greater part of the material. Agriculturally the Shropshire
area differs a good deal from most of the others, being generally
at a lower elevation and somewhat more fertile, partly, no doubt,
owing to the mixed nature of the soils and the presence of much
rainwash. The occurrence of numerous limestone beds also
has an ameliorating influence on the soils. In many places a
good deal of material has been washed down from Old Red
Sandstone outcrops at higher levels, and, as will appear in the
next chapter, this formation yields soils of remarkable fertility.
The Wenlock limestone forms steep ridges covered by woods,
but the Wenlock shales form a large expanse of land of somewhat
variable character; some parts are heavy clays, good for wheat

and beans. Taken as a whole, this formation is not very fertile, though there is some fairly good grass. The soils of the Ludlow beds, though still on the heavy side, are more loamy than those of the Wenlock shale. In the valleys is some good grass land, but the upland region of Clun Forest is open sheep walks, with a local breed of sheep.

The major part of the Silurian area of central and north Wales consists of high mountains and tablelands, with deep valleys between. On the hills the soils are largely peaty and covered with poor grass or heather, devoted to sheep. In the larger valleys there is a fair amount of arable land of mediocre quality and much coarse wet pasture. The Silurian tract of the Lake District covers a great area; on the whole it is less elevated and rugged than the part occupied by the Ordovician system, but still high and for the most part unenclosed ground, largely occupied by the Herdwick sheep. The character of the agriculture of Westmorland can be gathered from the fact that in 1907 there were only 127 acres of wheat in the whole county, while oats occupied nearly 14,000 acres. The total area of the county is 505,000 acres, hence the proportion of arable land is very small.

The Ordovician and Silurian rocks of the Southern Uplands of Scotland form a tract of ground for the most part at high elevations and largely devoted to sheep-farming. Though less rugged than the Lake District the country is yet largely uncultivated, being covered with short grass or with heather. Only in the valleys of the numerous rivers is there much arable land, but here the soil is often fertile and well-farmed. Here again however the subsoil is largely composed of drift and alluvium. The sheep-farming of the Southern Uplands is of great importance and has given rise to a considerable woollen industry at Galashiels and Selkirk. Further west in Ayrshire and Galloway the country is less elevated and forms a good agricultural district, where dairy-farming is largely carried on. The moist climate here is specially favourable to turnips and very heavy crops are grown.

Summary. With few and not very important exceptions the Precambrian and Lower Palaeozoic rocks of the British

Isles form mountainous and hilly ground in the northern and
western parts of the country. This disposition of the rocks is
a result of the general geological structure of the country. The
older rocks have undergone the highest degree of metamorphism,
being in consequence hard and resistant; also they have at
various times been folded and elevated into mountain chains.
The numerous intercalations of igneous rock are also hard and
denuded with difficulty. Consequently the older rocks form
the highest parts of the country. Again, climate has a most
important influence in determining the character of the soil and
vegetation. Mainly as a result of the present relief of the land
the climate of the British Isles shows extraordinary variations
for so small an area. This is clearly manifested by the rainfall
maps and by the peculiar distribution of the isotherms. The
peculiarities of the climate are of course greatly intensified by
the fact of the high land being in the west. If the west coast
were occupied by low land it would naturally be warm and damp
owing to the prevailing Atlantic winds bringing rain. The
effect of the high elevation of the western side is of course to
exaggerate the rainfall, while the effect of the higher temperature
is to a great extent neutralized by the height of the land. At the
sea-coast and in sheltered valleys in the west the climate is
surprisingly mild and vegetation luxuriant, but on the wind-
swept tablelands and on exposed portions of the mountains the
land is often almost barren, or only fit for the growth of heather
and peat at quite small elevations. Again, the comparatively
recent occurrence of glacial conditions among the mountains of
the British Isles has had an important influence on the soils.
The ice swept the weathered material from the hills, leaving in
many places nothing but bare unweathered rock. After the
departure of the ice, the conditions were such as favoured the
accumulation of peat rather than a renewed formation of
fertile agricultural soils, and in the Highlands of Scotland, the
western Isles, some parts of the Pennine Chain, in Wales and
over a large part of Ireland, the peaty condition has persisted to
the present day. Furthermore the glaciers laid down in the
valleys and on the low ground at the foot of the mountains
vast accumulations of boulder-clay, sands and gravels, consisting

of material largely in an unweathered condition and containing little plant food in the available state. Many of the soils on these glacial deposits, though potentially rich, are still somewhat raw and incomplete, since scarcely sufficient time has elapsed since the departure of the ice for the processes of chemical and bacterial decomposition to produce their full effect.

Again it must be admitted that the majority of the rocks composing the older formations were not such as would, even under favourable conditions, produce soils of first-class quality. The prevailing rock-types are slates and grits, neither of which usually yield fertile soils, even in more recent formations in better climates. There is commonly a deficiency of lime, and phosphoric acid may also be scarce. Potash is commonly present in sufficient quantity, but often in forms not readily available. As a result of all these causes working together, the soils of the older rocks are generally far from fertile and agriculture is, with few exceptions, in a backward state.

Besides all these natural factors there is also another influence to be reckoned with, namely, the character of the inhabitants of these districts. The Highlands of Scotland, Wales and Ireland are mainly inhabited by a race different from that dwelling in England and the Lowlands of Scotland, possessing different standards and ideals. The question is largely an economic one. Till recently these people were less advanced in civilization than the rest of the country; they possessed little capital and had made small progress in the application of modern methods to agriculture. It is an interesting, but perhaps somewhat unprofitable subject for discussion, to what extent this backwardness was due to the character of the people themselves, and to what extent to the natural disadvantages of climate and soil with which they had to contend. It may at any rate be admitted that in the latter respect they were heavily handicapped by Nature. The earlier formations also are singularly lacking in products of economic value, so that there was little incentive to migration from elsewhere into the districts occupied by them, leading to a growth of population round any particular centres. The inhabitants consequently remained in the pastoral stage of civilization, a state of affairs still prevailing over the greater part of the area. Of late years, owing to emigration, the population of certain large parts of Ireland and Scotland has greatly diminished and a good deal of land has actually gone out of cultivation.

From a consideration of the facts thus briefly outlined, it is clear that the ultimate basis of the social and economic problems of these regions is mainly geological.

CHAPTER XI

THE DEVONIAN AND CARBONIFEROUS SYSTEMS

Each of these systems covers a very large area in the British Isles and both are of great agricultural interest; the rocks of the Devonian system yield some of the best land in this country, and the Carboniferous, though not naturally very fertile, is nevertheless of great practical importance, largely owing to the coal contained in its upper part. The coal-fields are naturally centres of crowded population, and agriculture on them and in the neighbouring areas shows some special features.

I. THE DEVONIAN AND OLD RED SANDSTONE SYSTEM

One of the most remarkable features of this system is the occurrence of two different and sharply contrasted facies, varying widely from each other in almost all respects. These facies are known as the Devonian and Old Red Sandstone types respectively. The Devonian rocks in the strict sense of the term are in this country confined to the peninsula of Devon and Cornwall; the Old Red Sandstone facies is found in South Wales and the west of England, and very largely in Scotland, extending also into the north of Ireland. In the south-west of Ireland the rocks of this age again differ a good deal from both types, but show more resemblance to those of North Devon than to the Old Red Sandstone. These variations are to be accounted for by the occurrence in late Silurian and Devonian times of a great series of earth-movements, leading to profound modifications in the distribution of land and sea. The marine phase of the Lower Palaeozoic era came to an end, and over the greater part

of the British Isles continental conditions set in; only in Devonshire and Cornwall did open sea continue to prevail. This
great change was heralded by the gradual appearance of more
and more continental characters in the uppermost beds of the
Silurian. The Downton sandstone clearly shows an approximation to the Old Red Sandstone facies, both in its lithological
character and in its fossils. In Scotland volcanic rocks are
abundant in the Old Red Sandstone and the earth-movements
were accompanied by intrusion of great masses of granite in
northern England, Scotland and Ireland. Great mountain
ranges were formed from the older rocks, and their denudation
gave rise to the thick accumulations of the Old Red Sandstone
in the northern and western parts of the British Isles.

A.　The Marine Devonian Facies

The effects of these earth-movements were not conspicuous
south of the line now occupied by the Bristol Channel, and in
Devon and Cornwall marine conditions still prevailed. From
the varying character of the rocks in this area it can be shown
that the shore of this sea lay not far north of Devonshire,
while in South Devon and Cornwall the water was deeper.
The open sea was clearly limited also towards the east, since in
the Mendip Hills in Somerset the Old Red Sandstone facies is
seen. The Devonian rocks of north Devon and west Somerset
form a thick series of marine and often fossiliferous strata of
somewhat variable character. The base of the series is nowhere
visible and it is not known on what they rest. The greater
part of the succession consists of yellow, brown and reddish
sandstones and grits, but there are also important beds of slate
and limestone. In some respects the sandstones show a
considerable resemblance to the Old Red Sandstone in neighbouring areas, but they are shown by fossil evidence to be of
marine origin. The rocks have been divided into Lower,
Middle and Upper series, but much controversy has arisen as
to the true age of a part of the series known as the Morte Slates.
From their apparent stratigraphical position they should be
Middle Devonian, but the few and badly preserved fossils
prove them to be Lower Devonian or even older; it is

now generally believed that their present position is due to faulting.

The Devonian rocks of the northern type occur only in north Devon and west Somerset, occupying the area bounded on the north and west by the coast, on the south by the Taunton and Barnstaple railway to a point about 6 miles west of Taunton, the boundary then running north and meeting the coast again near Minehead. This area is generally high ground and includes the well-known Exmoor Forest, of which the highest point is Dunkery Beacon, 1707 feet. The eastern portion forms the Brendon Hills. The whole area is notorious for the steepness of its roads, and it is famous also as the home of the wild red deer. The highest ground is open and unenclosed, being covered with heather and gorse. In the steep narrow valleys are many woods, and the soils of the cultivated portions are not as a rule of first-class quality. The soils formed from the grits are often poor and stony, but the bands of slate and limestone are of better quality and more fertile.

The Devonian rocks of south Devon are separated from the area just described by a broad belt occupied by Carboniferous rocks, forming the middle part of a great syncline. Below this Carboniferous belt the Devonian rocks undergo a lithological change, emerging on the southern side as a series mainly argillaceous and calcareous, with well-cleaved slates and thick massive limestones. There is some uncertainty as to the base of this series, and of late years it has been shown that certain beds in southern Cornwall formerly believed to be Devonian are really Ordovician and perhaps also Silurian. So far as the sequence is known the lower part appears to be mainly composed of grits, while the Middle Devonian is more slaty, and the Upper Devonian mainly massive limestones. In Cornwall almost the whole series is slaty, and is generally known locally as "killas." The Devonian rocks have undergone intense metamorphism, and the structure of the region is very complicated in detail, owing to folding, faulting, overthrusting and igneous intrusions.

The most striking geological feature in Devon and Cornwall is the great granite intrusions, of post-Carboniferous age. The largest of these is the Dartmoor granite; this was intruded along the line of junction between the Devonian and Carboniferous rocks and forms a somewhat irregular mass about 20 miles in diameter. Dartmoor is a wild upland region, mainly uncultivated and covered with heather. There are valuable granite quarries and an important tin-mining industry; the latter however is gradually decreasing owing to exhaustion of the accessible ore. In Cornwall also there are four large granite masses and some smaller ones. The mining industry of southern Cornwall is still of great importance, as there are abundant veins containing tin, copper, zinc, arsenic and other ores. The total number of minerals found in the Cornish veins is very great and many are of great scientific interest as well as of practical value. Among some recent developments mention may be made of the exploitation of pitchblende as a source of radium. The granite intrusions have produced a high grade of metamorphism in the surrounding sediments and have themselves undergone some remarkable changes as a result of the action of highly heated vapours (pneumatolysis). These changes are of several kinds, but the most important economically is kaolinization, or the conversion of the felspar of the granite into china-clay[1]. This is the basis of an industry of great importance.

The relief of the area occupied by Devonian rocks is very varied, but on the whole the country is composed of a deeply dissected plateau, with steep-sided valleys; in these there is much woodland. For the most part the soils on the Devonian rocks are not naturally very fertile, but there is much rainwash and alluvium of good quality, and there are many prosperous agricultural districts. In many parts of Devon and Cornwall dairy-farming is of much importance. The volcanic rocks yield the richest soils; the slaty members of the Devonian system produce rather heavy clay soils, those on the grits being somewhat lighter in character. The limestone soils also are for the most part rather thin and clayey, and consequently much affected in seasons of drought. Owing to the mild climate

[1] Reid and Flett, "The Geology of the Land's End District," *Mem. Geol. Surv.* 1907, pp. 53–60.

many semitropical plants can be grown in the open air in sheltered valleys and near the coast, and the growing of early flowers and vegetables forms a very flourishing industry.

B. The Old Red Sandstone

The strata comprised under the general name of the Old Red Sandstone form a typical example of a *continental facies*. They include sandstones, shales, marls and conglomerates with occasional limestones; the fauna is very scanty and almost confined to fish; these however are abundant in some localities. In Scotland volcanic rocks are largely developed. The exact nature of the conditions of deposition of the Old Red Sandstone are not known with certainty, but it is at any rate clear that there were several disconnected areas. Some authorities maintain that these were lakes, while others hold them to have been long narrow arms of the sea, lying between the folded mountain chains of older rocks that began to arise towards the end of Silurian times. These (the Caledonian folds) had a general trend from S.W. to N.E., and were accompanied by violent faulting and overthrusting, together with intense dynamic metamorphism. From this time dates the greater part of the highly complex structure of the Scottish Highlands, as well as the present disposition of the older rocks of the Lake District, north and central Wales and a large part of Ireland. Earth-movements on so large a scale are always accompanied by uplift and great denudation, especially from the tops of the folds. The loose material thus formed is accumulated in the intervening hollows and gives rise to continental deposits, such as the Old Red Sandstone. Although, as above stated, the exact nature of these basins of deposition is not known, they have for convenience been called lakes, and a name has been assigned to each.

The Welsh Lake. The Old Red Sandstone rocks are exposed at the surface over a large area in the border counties and in South Wales. They possess a total thickness of several thousand feet, the lower part consisting of red and variegated sandstones and marls with occasional thin bands of limestone, known locally as "cornstones." The upper part is formed chiefly of red, brown and yellow sandstones and conglomerates, the latter being largely formed of pebbles of white quartz. The divisions usually adopted are as follows:

Upper Division ... Yellow and red sandstones and conglomerates, 500 feet.

Middle Division ... Brown sandstones (Brownstone series), 500 to 1500 feet.

Lower Division ... Cornstone series, 3000 to 4000 feet.

The only fossils are fish, such as *Cephalaspis*, *Pteraspis*, *Scaphaspis*, and near the top a very few examples of *Conularia*, a marine shell, show an occasional and limited connexion with the sea.

The Old Red Sandstone of this area yields soils of extraordinary fertility. The sandstones and marls give rise to red and brown soils of great depth, which are very largely under permanent pasture, affording probably the richest grass land in the whole of the British Isles; this is the home of the far-famed Hereford cattle. It is generally conceded that the Teme valley in Herefordshire contains the most fertile soils of this country, and both here and in Worcestershire there are extensive hop-gardens, comprising about one-third of the total area under this crop. Throughout the lower Severn valley there are extensive orchards of cider-apples. The Cornstones yield soils of a more brashy nature, better suited for corn growing; hence the name. On some of the high ground occupied by the uppermost division of the Old Red Sandstone, and especially on the conglomerates, the soils are of poor quality, being very stony. The great elevation is also unfavourable to high-class farming, especially in the Brecknock Beacons. The lower Severn valley is on the whole a district of high rents and high farming, contrasting strongly with the areas of Lower Palaeozoic rocks to the west. There are however even here occasional patches of poor land where the underlying rocks consist of nearly pure quartz sand, without the admixture of other decomposed minerals that contribute so much to general fertility.

The most northerly exposure of the Old Red Sandstone in this area is a few miles north of Bridgnorth, in Shropshire.

Between this point and the Scottish border the formation is only represented by a few scattered patches of conglomerate in west Yorkshire and Westmorland, and even these have been referred to the base of the Carboniferous. They probably belong to the Old Red Sandstone, but are of little agricultural importance, as they cover only a very small area.

Further north the Old Red Sandstone is clearly divisible into two series, Upper and Lower, there being usually a strongly-marked unconformity between them; indeed in parts of Scotland horizontal strata of the Upper series can be seen resting on an eroded surface formed by almost vertical beds of the Lower series. It is however by no means clear whether these correspond in time to the upper and lower divisions of the Welsh borderland. Some writers regard the "Upper Old Red Sandstone" of Scotland as really forming the base of the Carboniferous; this question is not yet settled.

In the Cheviot area both series are represented; the older consists partly of red felspathic sandstones and marls, but the greater part of the Cheviot Hills is formed by igneous rocks, including both lavas and ashes and a large boss of intrusive granite. The volcanic rocks are 2000 feet thick, and include several varieties of andesite and andesitic ash. The Cheviot Hills form an undulating upland region, rising to the height of about 2600 feet, and chiefly covered with short fine grass; this is a great sheep-farming district and is the original home of the Cheviot sheep, which have now spread widely over the Southern Uplands. The red sandstones and marls form rich agricultural land in Berwickshire and Roxburghshire, especially in the Teviot valley. These belong mainly to the Upper Old Red Sandstone, forming the so-called Tuedian series; though lithologically of Old Red Sandstone type, they may be wholly of Carboniferous age. Their outcrop extends continuously into East Lothian, there forming some of the most fertile and most highly rented land in the whole of the British Isles. The soil is specially suited to the growth of potatoes of fine quality. The great fertility of the soil is no doubt largely attributable to the presence of much volcanic material in the rocks, which by its gradual decomposition supplies large quantities of plant food.

Old Red Sandstone rocks form the surface over a considerable area in the central valley of Scotland. In its general structure this area is a syncline, the older rocks cropping out along the northern and southern edges, while the middle part is occupied by Carboniferous strata. Both on the north and on the south the valley is bounded by faults, so that in its structure it is essentially a "rift-valley." The sediments of Old Red Sandstone age attain a great thickness, and there is in addition a large development of volcanic rocks, both lavas and ashes. The latter attain a thickness of some 6000 feet in the Ochil and Sidlaw Hills on the north, and appear also in the Pentland Hills south of Edinburgh. The volcanic rocks are confined to the lower division of the system.

The Lower Old Red Sandstone of Forfar, which probably attains a total thickness of 18,000 or 20,000 feet, consists of a variable series of sandstones, conglomerates, flagstones and shales, with remains of fish and plants. In the Ochil Hills the conglomerates and breccias contain many blocks of volcanic rock, indicating the existence of volcanoes in the near neighbourhood. The prevailing colour of the sediments is red, but some of the flags and shales are grey. The volcanic rocks are chiefly andesitic in character, being very similar to those of the Cheviot Hills.

The Upper Old Red Sandstone, lying unconformably on the Lower division, consists of conglomerates and red sandstones below, with occasional beds of limestone, the upper part being yellow sandstones and shales. The thickness is about 2000 feet. The uppermost division contains many fish-remains, the most famous locality being at Dura Den, near Cupar.

The Old Red Sandstone forms a broad tract, extending from Helensburgh and Dumbarton on the Clyde, to the east coast at Stonehaven, and having a maximum width of some 25 miles. This includes the well-known agricultural districts of Strathmore and the Carse of Gowrie, comprising some of the most highly farmed land in the British Isles. The fertility of the soil is no doubt largely due to the presence of much volcanic material, just as in the case of the Lothians. In many parts however the actual soil is formed from boulder-clay and other glacial deposits.

Scattered patches of Old Red Sandstone sediments occur in several places in the central and southern Highlands, showing that the strata once extended further than they do now, having been largely removed by denudation. In the western Highlands, in the neighbourhood of Oban and Glencoe, there is also a considerable area of rocks of the same age, largely volcanic, but with some remarkably coarse conglomerates. None of these are of any agricultural importance.

Along both shores of the Moray Firth, in Caithness and in the Orkneys, Old Red Sandstone strata cover a considerable area. They seem to belong to the lower division of the series, but are of a type somewhat different from that found in other parts of Scotland. Red sandstones and conglomerates are still seen, especially near the base, but the main mass consists of shales and flagstones, the whole group being known collectively as the Caithness Flags. The total thickness is about 16,000 feet. Fossil fish are very abundant. The upper division is scantily represented by a few hundred feet of yellow and red sandstones in Moray and Nairn, with another small patch near Dornoch. There is as usual a marked unconformity at the base.

The soils yielded by the Old Red Sandstone formation in this area are again of great fertility and this is all the more remarkable in view of the comparatively high latitude, 57° to 59° N. This applies with almost equal force to all the Scottish areas of Old Red Sandstone, all those on the eastern side of the country being districts of great fertility, with high farming. This state of things is due to a combination of several factors, the chief being (a) the inherent nature of the rocks, (b) the comparatively low elevation, and (c) the climate. The presence of a thick covering of drift over most of the low ground has also an important influence in modifying the character of the soil, by mixing together materials of varying composition and thus supplying all the elements of plant food, often in a partially decomposed and readily available form.

The Old Red Sandstone sediments were largely formed by denudation of crystalline rocks in a dry climate, so that there was little removal of soluble matter in solution. There was also in many places a considerable admixture of contemporaneous volcanic material and, as is well known, the decomposition of lava nearly always yields a rich soil. The rocks are also of such a nature as to yield naturally a deep, free-working and well-drained soil, easy to cultivate and easily penetrated by the roots of plants. Thus all the natural elements of fertility are present. But these alone are not sufficient to give rise to good land unless other conditions, especially the climate, are favourable. To show this we have only to turn to the tracts covered by the Torridon Sandstone along the west coast of Sutherland, Ross and Cromarty, in the same latitude as the shores of the Moray Firth. This formation is lithologically very like the Old Red Sandstone and was indeed once believed to form part of it, though this is now known to be erroneous. The Torridon Sandstone forms high, bleak mountains and plateaux, covered with heather and peat, and hardly cultivated. The climate, though not really very cold, is wet and sunless, and corn can scarcely be got to ripen even at sea-level in this district. The only corn crop cultivated to any extent is oats.

But when we turn to the Old Red Sandstone tract of Caithness a remarkable contrast is seen. The land is well suited to the growth of roots and barley, and distilling is therefore an important industry, while cattle feeding is carried on extensively. The lower elevation of the ground certainly has a considerable influence here, but it appears that the most important factor is climatic, namely, the smaller rainfall with correspondingly greater amount of sunshine. Extremes of climate are not in themselves inimical to the growth of good crops, as shown by the great fertility of parts of the prairie region of America. The long cold winter is counterbalanced by the greater heat and sunshine of summer, and the same applies in a lesser degree to the east coast of Scotland, where the growing season, though short, is sunny, and the summer days are very long. Nevertheless it is clear that the character of the soils themselves is of the greatest importance in determining their fertility and agricultural value. Good climatic conditions alone will not make all soils equally fertile, though they enhance greatly the value of a naturally good soil, such as is yielded by the Old Red Sandstone of Scotland.

II. The Carboniferous System

Where the Upper Old Red Sandstone is present it is commonly overlain conformably by the Carboniferous; indeed, as before stated, the actual line of demarcation between the two systems is in many places quite uncertain. But the Carboniferous system overlaps the Old Red Sandstone, and over very large areas rests unconformably on older rocks, either Lower Palaeozoic or even Precambrian. The rocks of the Carboniferous form a larger proportion of the surface of the British Isles than any other formation, and in addition their economic products are of greater commercial importance than those of any other system. The coal-fields are all regions of concentrated industrial activity and thickly populated; hence the agricultural conditions are of a somewhat special character.

The rocks of the Carboniferous system show considerable variation, both vertically and horizontally. The most constant features are the preponderance of limestones in the lower part and of sandstones and shales with coal in the upper part, though even in these respects there are local variations of facies. For example, the Carboniferous rocks of Devonshire belong to a special facies, the Culm type, unknown elsewhere in Britain, though recurring on the continent. Again in Scotland the coal-measure facies extends into the lower part of the system. These variations of facies are due to the geographical conditions prevailing at the time of formation. At the end of Old Red Sandstone times the British area was subjected to a general tilting movement, producing fairly deep sea in the south, gradually shallowing, with local irregularities, towards the north. Beyond the present southern border of the Highlands there was a great northern continent, the source of much of the sedimentary material. As time went on this difference was gradually obliterated and at the end of the period the whole area south of the Highlands consisted of low-lying land and shallow water, with abundant swamps and lagoons, giving rise to the conditions necessary for coal-formation. The highest Carboniferous strata give indications of the approach of a continental phase, which culminated in the succeeding Permian and Triassic periods.

The Carboniferous system is divided into two main series, the Lower and Upper Carboniferous; these in most localities differ so much that it is convenient to treat each separately. This plan greatly facilitates comparison of different areas.

A. The Lower Carboniferous

The Lower Carboniferous strata of Devonshire consist of a very thin series of peculiar character, chiefly composed of black limestones and cherts, the whole probably formed in deep water. Since the total thickness is only a few hundred feet and the strata are highly inclined, the outcrop forms only a very narrow band along each side of the Devon syncline, and the series is of no agricultural importance. The chief fossils are *Posidonomya Becheri* and Goniatites (a kind of primitive Ammonite).

Lower Carboniferous rocks outcrop at the surface over large areas in Somerset, Gloucestershire and South Wales, and their development is very complete; in fact the Bristol district must now be regarded as the type area for this formation. The typical section is that so well seen in the Avon gorge between Bristol and the sea. The Lower Carboniferous has here been divided into zones characterized by corals and brachiopods; these zones are of wide application elsewhere in this country. The total thickness is a little over 2000 feet and the rocks were originally divided into the Lower Limestone Shales below (350 feet) and the Carboniferous Limestone proper above. These divisions, though unscientific, are practically useful, since they correspond to variations of lithological character. The main mass consists of highly fossiliferous marine limestone of varying texture and colour, with occasional shale bands. The general colour is some shade of grey or even black and the limestones are very massive, consisting largely of broken shells and corals, with Foraminifera. The corals, *Lithostrotion* and *Syringopora*,

are common, with species of *Productus, Spirifer* and allied
genera of brachiopods. The highest beds are shaly and sandy,
passing up gradually into the overlying Millstone Grit. The
Carboniferous Limestone also forms the greater part of the
Mendip Hills, where its characteristic features are particularly
well seen at Cheddar. North of Bristol the series becomes
much thinner; round the Forest of Dean coalfield it is only
about 600 feet thick and near Newport it appears to be reduced
to 350 feet, but thickens again westward. All over this area
the dips are as a rule fairly steep, so that outcrops are narrow.
In the southern part of Pembrokeshire, near Tenby, the strata
are in many places vertical.

In several of the Midland coal-fields Lower Carboniferous
strata are completely absent, the Coal-measures resting directly
on Silurian, Cambrian or Precambrian rocks. This area
must have been land during Lower Carboniferous times, but
it is uncertain whether it was an island, or a peninsula extending
eastwards from high land in Wales. Lower Carboniferous
rocks again appear in the north of Shropshire, and reach a
considerable development in Flint and Denbigh. There are
also a few scattered patches of the limestone in the north of
Leicestershire. But in Derbyshire we find the southern limit
of the largest and most important outcrop of Carboniferous rocks
in the British Isles, namely, that of the Pennine range. From
a little north of Derby Carboniferous rocks extend uninterrupt-
edly to beyond the Scottish border and the width of the outcrop
is considerable. Between Derbyshire and the Border the rocks
undergo a good deal of lateral variation; the general character
of the change can perhaps best be summed up by saying that
the further north we go the lower does the shallow water and
terrestrial facies descend in the series. The massive marine
limestones of the south are gradually replaced by shales and
grits with thin impure limestones, indicating a shallowing of
the water towards the north. In the north of Northumberland
and in Berwick the total amount of limestone is very small. In
Derbyshire the base of the system is not seen and the total
thickness of the Carboniferous Limestone is unknown, but it
is noteworthy that though the visible thickness is some 2500 or

even 3000 feet, the whole is shown by its fossils to belong only to the uppermost zones of the Bristol area. In west Yorkshire, near Settle and Ingleton, the base of the Carboniferous is seen, resting on Silurian and older rocks. The base of the series here occurs at a horizon roughly equivalent to the middle of the Bristol succession, whereas in Westmorland most of the series is again represented. Hence it is clear that the Carboniferous Limestone was deposited on an uneven surface and that deposition began in different regions at widely varying times, as the ancient land was denuded away or submerged; in the Midlands deposition did not commence at all till Upper Carboniferous times.

In west Yorkshire the Carboniferous rocks are affected by some important faults, the most noteworthy being the Craven faults, near Ingleton and Settle. It is remarkable that the development of the strata shows conspicuous differences on either side of the faults, suggesting that even at that time some kind of natural barrier existed. South of the Craven Faults, near Clitheroe, the Lower Carboniferous series, which is highly folded, is over 3200 feet thick. North of the fault at Ingleborough the maximum thickness of the same strata is not more than 1400 feet, and the lithological character is different. This point will be referred to again in connexion with the Millstone Grit. In Westmorland and Cumberland the limestone rests on a considerable thickness of very coarse red conglomerate, which is probably of Old Red Sandstone age. In many places, e.g. in the Cross Fell escarpment, the base of the undoubted Carboniferous is formed by a conglomeratic sandstone; above this comes the main mass of limestone, the Melmerby Scar limestone.

Throughout the Pennine range the uppermost part of the Lower Carboniferous is formed by a series of shales and grits with thin limestones, known collectively as the Yoredale series. These beds show the oncoming of Upper Carboniferous conditions.

When the Lower Carboniferous is followed into Scotland the lateral variations of facies are found to be very strongly marked. The lowest division near the Border consists of some 2000 feet

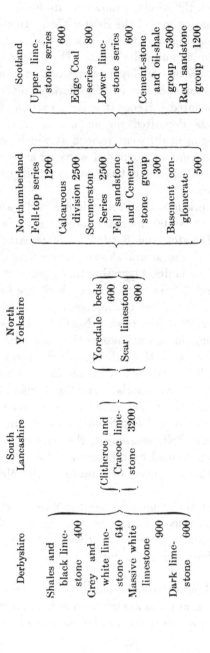

Derbyshire	South Lancashire	North Yorkshire	Northumberland	Scotland
Shales and black limestone 400			Fell-top series 1200	Upper limestone series 600
Grey and white limestone 640	Clitheroe and Cracoe limestone 3200	Yoredale beds 600	Calcareous division 2500	Edge Coal series 800
Massive white limestone 900		Scar limestone 800	Scremerston Series 2500	Lower limestone series 600
Dark limestone 600			Fell sandstone and Cement-stone group 300	Cement-stone and oil-shale group 5300
			Basement conglomerate 500	Red sandstone group 1200

of red sandstone, named the Tuedian series, from the river Tweed, while the calcareous facies has completely disappeared, being replaced by rocks like the Yoredales of Yorkshire. In the central valley of Scotland the more important of the coal seams are found in the upper part of the Lower Carboniferous, the Edge Coal group, while below this come the Cement-stone and Oil-shale groups, passing down conformably into the Upper Old Red Sandstone.

The succession and equivalence of the Lower Carboniferous rocks of the Pennine area and Scotland can be most clearly shown in a tabular form, as on the opposite page.

From the above descriptions it is clear that the rocks of this series show wide variations of character, and will therefore yield very varying soils. This subject can be most profitably discussed after the Millstone Grit has been described, since the two divisions are intimately connected; this course will avoid some repetition, otherwise inevitable.

B. The Millstone Grit

Almost everywhere the lowest member of the Upper Carboniferous series is made up to a greater or less extent of sandstones or grits, of coarse texture and sometimes conglomeratic in their nature. Intercalated with beds of this type are shales and occasional limestones. The general name of the series is due to the fact that some of the grits were found specially suitable for use as millstones, owing to their very rough and gritty texture. The Millstone Grit type of deposit is strictly a facies rather than a definite time-division, and it is quite clear that both the top and the bottom of the division as at present defined are at different horizons in different localities. Hence the demarcation of the Lower and Upper Carboniferous presents some difficulties. For example, the Millstone Grit of Bristol is certainly represented in part by the uppermost zone of the Carboniferous Limestone of Flint and Denbigh. However for our present purpose the lithological change is of the most importance, and the term Millstone Grit will here be employed in a somewhat old-fashioned sense.

Near Bristol and in South Wales the strata between the top of the limestone and the base of the Coal-measures consist of about 450 feet of yellow sandstone, usually of rather fine texture. Over most of the Midlands this division is absent, but in Denbigh, Cheshire and Derbyshire it sets in again and thickens rapidly towards the north, reaching a total thickness of perhaps 4000 feet in east Lancashire. In this area there are now recognized two series, the Pendleside group below, and the Millstone Grit above. The Pendleside group consists chiefly of shales and black limestones, and was formerly regarded as equivalent to the Yoredale beds of Yorkshire and therefore of Lower Carboniferous age; it is now recognized from its fossils to belong to the Upper Carboniferous. The Millstone Grit proper consists of an alternation of massive grits and black shales. The grits often form conspicuous escarpments and table-lands, such as Kinder Scout and Blackstone Edge. North of Settle the Pendleside group cannot be identified, and the Millstone Grit dwindles down to 600 feet in Yoredale, though it is said to thicken again in Durham and Northumberland, where it contains several workable coal-seams. In Scotland the Millstone Grit is represented by the "Moorstone Rock," or Rosslyn sandstone, from 350 to 600 feet thick, near Edinburgh, but thinning rapidly to the west.

From the foregoing brief account it is clear that the Millstone Grit is a variable formation, including rocks of many different types, likely to yield soils of very varying character. It appears to be in the main a shallow-water marine facies, foreshadowing the oncoming of the continental conditions of the Coal-measures.

The calcareous facies of the Lower Carboniferous is one of the most striking rock-types to be found in this country. Where present, in Great Britain, it commonly forms regions of considerable elevation, such as the Mendips and the hills of Derbyshire and west Yorkshire. Hence it was formerly known as the Mountain Limestone, a useful name now generally abandoned. It is only in Ireland that this limestone occurs to a large extent in low ground, forming the substratum of a great proportion of the boggy plain in the centre of that country.

The Carboniferous Limestone shows very perfectly the peculiar kind of denudation characteristic of calcareous rocks, now generally designated as the Karst type, from a district on the eastern side of the Adriatic (see p. 71). Where Carboniferous Limestone forms a level area, free from drift, it usually bears no soil, the surface consisting of bare rock, with many and deep open fissures called "grikes" in Yorkshire; in these alone does any vegetation manage to exist. This state of affairs is well seen on the "clints" of Ingleborough and near Grange in Westmorland. Such bare rock surfaces are common in all Carboniferous Limestone districts; other conspicuous features are steep escarpments, dry valleys, rock-pillars, caves, swallow-holes and underground rivers. All of these are well seen in Derbyshire and west Yorkshire. Where some soil does exist on the limestone it is generally covered with short, sweet turf, heather and bracken being almost unknown even at high elevations. In parts of Yorkshire the limestone is masked by drift; under these circumstances the character of the soil and vegetation is modified and peat may be formed. Even then however the underground drainage has an important influence on the superficial formations, and the surface often shows many funnel-shaped hollows, due to the existence of swallow-holes below.

The Lower Carboniferous rocks and the Millstone Grit of the Pennine Chain form a great mass of high ground, dissected by deep valleys and often rising into mountains, as in Derbyshire and west Yorkshire. On the borders of Westmorland, Durham, Cumberland and Northumberland is a great plateau over 2000 feet in height, the highest point, Cross Fell, nearly attaining 3000 feet. In the southern part of the range limestones are dominant, showing the characteristic features before described, but further north the strata consist chiefly of an alternation of massive grits and shales, with only subordinate calcareous bands. The grits and shales of the higher regions are very largely open moorland, often very peaty, and covered with heather. To the south-west of the Craven fault is a large extent of comparatively low ground, composed of Millstone Grit, Pendleside and Yoredale beds. The latter division is also prominent on the eastern side of the watershed, and in Wensleydale it forms an area of fertile land, largely devoted to dairying (cheese-making). The soils derived from the grit rocks are often stony or light and sandy, but the shales are somewhat heavier and of better quality, though often in need of drainage.

C. The Coal-measures

From the economic standpoint this division of the Upper Carboniferous is the most important formation in Great Britain. It forms the surface of several very considerable areas, and its underground extension is also known to be great beneath newer rocks. With the buried coal-fields we are not here concerned, except in a very general way. The Coal-measures do not as a whole form very good agricultural land, but owing to the density of population those parts not actually built over are almost entirely under cultivation of one sort or another.

The Coal-measures may be regarded as a continuation and exaggeration of the shallow-water and continental facies of the Millstone Grit, and of the Lower Carboniferous of the northern areas. In fact, as already pointed out, Coal-measure conditions set in during the Lower Carboniferous period in Scotland. The Coal-measures consist of a very variable series of sandstones, clays and shales, with seams of coal and occasional bands of marine limestone, indicating temporary incursions of the sea. In parts of the Midlands, in Staffordshire and in Warwickshire, the uppermost beds are of a very different type, consisting of red sandstones, marls and conglomerates, with little or no coal, indicating the approach of a terrestrial phase.

The present disposition of the coal-fields of England and Wales is the result of earth-movements of the Armorican series, that took place after the Carboniferous period, but before the Permian. There were then formed two sets of folds, striking E.–W. and N.–S. respectively, the former being dominant in the south, the latter in the north. From the tops of the anticlines thus formed the Upper Carboniferous rocks were removed by denudation, the Coal-measures being preserved in the intervening basins.

The visible coal-fields of Great Britain may be conveniently divided into four groups, as follows:

1. South Wales, Forest of Dean and Bristol.
2. The Midlands.
3. The Pennines.
4. The central valley of Scotland.

There is a large area of Upper Carboniferous rocks forming the centre of the Devon syncline, but these belong to the Culm facies and do not contain workable coal-seams.

The Coal-measures of north Somerset do not form a large area at the surface, the coal being largely worked below a cover of Secondary rocks, but the South Wales coal-field is one of the largest in the country. It formed originally a high-lying plateau, now deeply dissected by numerous north and south valleys, such as Taff Vale and the Rhondda valley. The general structure is an oval basin, with its long axis running east and west, but there are several subsidiary folds, throwing the rocks into undulations. In Pembrokeshire the structure is more complex and folding and faulting more strongly marked.

Three divisions are generally recognized in the Coal-measures of this area:

Upper Coal-measures	...	3000 feet.
Pennant Grit	...	...	3000 ,,
Lower Coal-measures	...	1600 ,,

It is to be noted that the "Upper" Coal-measures of South Wales are equivalent to the upper part of the Middle Coal-measures of the Midland coal-fields. The Lower and Upper measures both contain abundant seams of coal, interstratified with shales and some sandstones, while the Pennant series is mainly composed of felspathic grits, with little or no coal, except in the west, near Swansea.

The Midland group of coal-fields includes several scattered patches of no great individual size, but collectively of much industrial importance. They are mainly situated in Warwickshire, Leicestershire, south Staffordshire and Shropshire. For the most part they rest directly on older rocks, the Lower

Carboniferous being absent. In all these areas the "productive" Coal-measures consist of grey or white sandstones, with good seams of coal, one seam in Warwickshire being 25 feet thick. The upper Coal-measures, where present, are usually red in colour, without coal, and a large part of them have been mapped as Permian by the Geological Survey.

The coal-fields of the Pennine group include the following:

North Staffordshire,
Lancashire,
Yorkshire, Derbyshire and Nottinghamshire,
Durham and Northumberland,
Cumberland.

The Denbigh and Flint coal-field is the western limb of a syncline whose eastern limb forms the north Staffordshire coal-field. The beds are no doubt continuous below the plain of Cheshire, though too deep in the centre to be workable.

The Coal-measures of north Staffordshire and Lancashire show a general succession as follows:

Upper Coal-measures	Red and purple sandstones and marls	2000 feet.
Middle Coal-measures	Grey shale and sandstones with coal and ironstone	3500 ,,
Lower Coal-measures	Grey flags and black shales with thin coals	1800 ,,

Workable coal-seams are almost confined to the middle division; in south Lancashire the total thickness of coal is from 70 to 80 feet, and workable beds of ironstone are abundant. The red measures are well developed in the Potteries coal-field in north Staffordshire, but they are not well seen in Lancashire. They contain no coal.

The Coal-measures form the surface rock over a large area in Nottinghamshire, Derbyshire and Yorkshire, and they are known to extend underground far to the east. The lower division both here and in Lancashire is often called the Gannister series, this being a miners' term for a hard white siliceous sandstone that often occurs below the coal-seams. The middle Coal-measures have a very great thickness and are closely similar to those of Lancashire, many of the seams being the same. The red measures are only known in borings in Nottinghamshire and near Rotherham.

The Gannister beds become very thin in Durham and Northumberland, probably not more than 150 feet, but the Middle Coal-measures are some 2000 feet thick, with about twenty seams of coal. It is not known whether any Upper Coal-measures are present in this area. The Cumberland coal-field consists only of a small patch near Whitehaven, the relic of a once much larger area that was probably destroyed during Tertiary times by the uplift of the Lake District dome.

The Upper Carboniferous series is not well developed in Scotland, where most of the productive coal-seams are in the lower division of the system. In the Clyde basin the Lower Coal-measures contain ten workable seams, with several important bands of ironstone. The Upper Coal-measures of Scotland are red and unproductive, and their relation to the strata of the English coal-fields is still uncertain; they rest unconformably on the lower division.

In Ireland Coal-measures are found in Kilkenny and Queen's County, in Leinster, where there are a few good seams of coal, and also in Clare, Limerick, Kerry and Cork, forming the Munster coal-field, where the seams are thin and variable. The rocks consist of sandstones, flags and shales and have a maximum thickness of about 3500 feet in Munster and 1700 feet in Leinster. Commercially they are unimportant.

The fossils of the Coal-measures differ greatly from those of the Carboniferous Limestone, belonging chiefly to those types characteristic of estuarine, lagoon and fresh-water conditions. By far the most important are remains of plants, together with fresh-water mollusca, fish, amphibia, crustacea and even insects. Here and there are found bands of limestone with a rather scanty marine fauna. These marine bands however are very useful for purposes of correlation and serve as indices in locating the horizons of the different coal-seams. The flora of the Coal Measures includes gigantic lycopods (club mosses) such as *Lepidodendron* and *Sigillaria*; the roots of the latter are often called *Stigmaria*. *Calamites* is a large species of Horse-tail (Equisetaceae). True conifers are also known, but some of the most abundant forms, formerly classed with the ferns, are now known to be intermediate between the ferns and the cycads; *Alethopteris, Neuropteris, Pecopteris* and others. Among the fresh-water mollusca the most important are *Carbonicola, Anthracomya* and *Naiadites*. The marine fauna includes many forms common also in the Millstone Grit and Pendleside series, some being found in the Lower Carboniferous. Some of the more important are Goniatites, *Bellerophon, Posidonomya Becheri, Aviculopecten papyraceus, Productus semireticulatus,* and *Spirifer Urei*.

Since the strata composing the Coal-measures show much variation of lithological character the soils naturally also vary very widely. The chief rock-types are sandstone, shale and clay, often occurring in comparatively thin beds in rapid alternation. In general terms it may be said that the sandstones yield light soils, the shales and clays heavy, cold soils, but both types agree in being on the whole of poor quality, of low natural fertility. Furthermore, over large areas the growth of crops is seriously hampered by the smoke from numerous industrial centres and factories, not only in towns, but scattered about the country districts. It is surprising over what wide areas this effect can be traced. Again in many places much land is spoilt by the flow of surface water contaminated with poisonous substances from pit-banks, slag-heaps and many other sources. Even if not directly poisonous to crops, these waters, which are often strongly acid, help to dissolve out the soluble plant food in the soil and especially to remove the lime, which as a rule is naturally deficient. Nevertheless, in spite of many drawbacks, owing to special geographical and economic causes, much profitable farming is found on the Coal-measures. It is however of a highly specialized type. As an example we may take the West Riding of Yorkshire, a typical industrial area supporting a vast population. In such an area, as is natural, the boundary between farming and market gardening is very difficult to draw, and certain crops, such as rhubarb, usually only found in gardens, are here cultivated as field crops. Another very special industry is the growing of green crops to be mown as fodder for pit ponies. As before stated, it is here possible to recognize two distinct types of soil, heavy and light, which are often sharply marked off over small areas. Arable land predominates, only the very heavy soils being under permanent pasture and meadow. In many parts, especially near the large towns, practically the whole produce of the land is sold off and great quantities of town manure are purchased to maintain a good standard of imparted fertility. Potatoes receive special attention and are grown with corn, mainly wheat, and clover, in a five-course rotation. In districts where there is sufficient grass the production of milk for sale is of considerable importance, especially near towns. In the more remote districts, away from railways, there is a more ordinary type of farming, but the district is not good for sheep, since the smoke is injurious to them; in the more populous districts the vast number of dogs kept by the inhabitants render sheep-farming a somewhat anxious business.

This description of one special district may be applied with little variation to most of the British coal-fields, where the industrial conditions rather than the actual character of the soil usually determine the type of farming. There are however some notable variations. The Pennant Grit of South Wales, which contains no workable coal, forms an elevated moorland tract, and in Gloucestershire there is a large area of woodland (the Forest of Dean). The uppermost red Coal-measures of the Midlands form soils very like those of the overlying Trias. In Devonshire there is a great area of sandstones and shales without coal (the Culm measures) that form poor wet clays, or stony sandy soils. There is a good deal of woodland on the sides of the steep valleys.

CHAPTER XII

The highest beds of the Carboniferous of the Midland coal-fields show a prevailing red colour, indicating the oncoming of a phase of continental conditions. This was the result of the great earth-movements of the Armorican series, which, as already mentioned, were responsible for the present arrangement of the Coal-measures in isolated basins. For a time denudation was dominant over deposition, removing the Carboniferous strata from the anticlines and depositing the material in the intervening synclines; consequently Permian strata occur in isolated patches of which only two now occupy any considerable area. During the Triassic period the arid continental type of deposition became much more widespread in Britain, and Triassic strata form a very considerable proportion of the surface, especially in the northern and western Midlands.

The unconformity between the Carboniferous and Permian strata is one of the most conspicuous in the British Isles and is seen with special clearness in Nottinghamshire, Yorkshire and Durham. This unconformity probably represents a considerable lapse of time, since the true upper Coal-measures (Stephanian) of France and Germany appear to be unrepresented in Britain. In India and in the southern hemisphere there is no break in the series, as for example in the Karroo system of South Africa and the Gondwana system of India, which range from Carboniferous to Trias.

In conformity with their origin during a period of strongly marked continental conditions the strata of the Permian and Triassic systems possess characters strongly resembling those of the deposits now being laid down in arid regions; in fact throughout both systems there is abundant evidence of the prevalence of desert conditions in this country. Marine strata, where developed at all, are peculiar, with a scanty fauna; beds of rock-salt, gypsum and dolomitic limestone occur on a large scale, and even the sandstones and marls show undoubted evidence of wind-action; in places the pitted, scored and polished surfaces of the older underlying rocks have been preserved to this day. In their general characters the Permian and Trias are very similar, and it has been proposed to class them all together, as the New Red Sandstone system. Lithologically and stratigraphically there is abundant justification for this, but unfortunately in their marine equivalents in other parts of the world, one of the most important of all palaeontological breaks occurs in the middle of the series; the Palaeozoic fauna, characterized by trilobites, here gives way to the Mesozoic ammonite fauna. Hence although there is little or no stratigraphical break, the dividing line between the Palaeozoic and Mesozoic epochs has been drawn at this horizon. As will appear later, the demarcation of Permian and Trias presents some difficulty in Britain and if this country alone were to be considered there would be no hesitation in the adoption of the New Red Sandstone system.

I. The Permian System

The Permian rocks of England form two principal areas, one on either side of the Pennine chain, with some smaller patches in the Midlands and in Devonshire; as before mentioned, most of the red rocks formerly mapped as Permian in the Midlands are now known to be Upper Carboniferous. The largest area of Permian rocks forms a narrow strip running in a nearly straight line from near Nottingham to the mouth of the Tyne. The two lower divisions, the Yellow sands and the Marl-slate, are quite unimportant, but the higher member, the Magnesian Limestone series, has a maximum thickness of about 800 feet in Durham, where it consists chiefly of dolomite rock, often with a brecciated or concretionary structure. Towards

the south it becomes thinner and beds of red sandstone and marl
come in; in Nottinghamshire about half of the series is marly
and sandy, and the visible thickness is small, owing to overlap
of the Trias. Fossils are scarce in the Magnesian Limestone, as
a result of the conversion of the original marine limestone into
dolomite-rock. The most common are *Productus horridus*,
Schizodus obscurus, *Fenestella retiformis* and fish (*Palaeoniscus*
and others).

The Permian rocks of the Eden valley in Westmorland and
Cumberland are very different in thickness and lithological
character, although the divisions can be correlated, as shown
in the following table:

Western basin			Eastern basin		
Magnesian Limestone		0–30 ft.	Magnesian Limestone		500–800 ft.
Hilton Shales	...	150 ,,	Marl Slate ...	...	3– 50 ,,
Penrith Sandstone ...		1500 ,,	Yellow Sands	...	0–100 ,,

The Penrith Sandstone affords a fine example of a true
desert deposit. The maximum thickness near Appleby is
about 1500 feet, consisting of bright red false-bedded sandstone,
with well-rounded grains and other indications of wind-transport. This affords an excellent building-stone. Near the base
and summit are beds of breccia, known locally as "brockrams."
These consist of angular fragments of Carboniferous Limestone
embedded in a red marly or fine sandy matrix. The fragments
appear to have been derived from the Cross Fell fault-scarp,
which must have been in process of formation during Permian
times. The brockrams are thickest in the south, near Kirkby
Stephen and disappear about Penrith. The Hilton Shales
contain plant remains like those of the Marl-slate. The so-
called Magnesian Limestone, really a sandstone or flagstone
with a dolomite cement, is quite insignificant.

There are very small patches of Permian sandstones at
Ingleton, Clitheroe and in the neighbourhood of Manchester,
but not much is known about them. In the Midland counties
the only undoubted Permian rocks are found in Shropshire
(Enville marls) and on the eastern flanks of the Malvern Hills.
These are not thick and cover only a very small area.

The lower part of the great series of red sandstones and conglomerates so conspicuous in Devonshire is considered to be of Permian age, but the line of demarcation from the Trias is somewhat uncertain. These rocks are splendidly exposed in the cliffs about Dawlish and Teignmouth, and again just beyond Torquay. They consist partly of bright brick-red sandstones and partly of conglomerates or breccias, composed of large and small fragments of older rocks of many kinds, embedded in a red sandy matrix. Near Exeter there are some igneous rocks, quartz-porphyry and basalt, with coarse breccias. The total thickness commonly assigned to the Permian system in Devonshire is about 2000 feet.

Since the subdivisions of the Permian rocks show so much variation of character they naturally yield soils of very different quality. The Penrith Sandstone, when not covered by drift, yields a very light soil, often woodland or heathy ground, but it is generally modified by a cover of boulder-clay. The soils of the eastern outcrop also vary a good deal, especially in Nottinghamshire, where beds of sandstone and marl partly replace the Magnesian Limestone. The Magnesian Limestone typically yields a brown loamy soil but is often covered by a stiff clay; this is possibly due to relics of shales or marls which formerly covered it, but have been mostly removed by denudation. The marls, when unmixed, form a cold tenacious soil, but when, as often happens, they are mixed with sands, the result is a deep fertile loam yielding good land. The lower mottled sandstones form a very light sandy soil. When followed northwards into Yorkshire the marls and sands gradually disappear, and the escarpment of the Magnesian Limestone forms a conspicuous plateau, declining gently to the east. This is very well seen about Ferrybridge. This area carries rather light soils, often thin and of somewhat poor quality; the most important crops are barley and turnips. Where the soil is rather more loamy, especially towards the borders of Durham, there is a larger proportion of grass land. The Permian area of south Durham is very largely covered by boulder-clay. A crop peculiar to the loams of the Magnesian Limestone is liquorice, which is grown near Pontefract.

II. The Triassic System.

The Triassic rocks of the British Isles cover a very large area, especially in the Midland counties, and constitute from the agricultural point of view one of the most important systems. As before remarked, it is often difficult to fix the dividing line between Permian and Trias, owing to lithological similarity and rarity or absence of fossils. Where the two formations are clearly seen in contact there is sometimes a slight unconformity, and the Trias overlaps the Permian in almost all directions; furthermore the upper divisions of the Trias overlap the lower divisions in many places, resting directly on Carboniferous and older rocks.

The Trias of Britain is divided, mainly on lithological grounds, into three groups, namely:

> Rhaetic series,
> Keuper series,
> Bunter series.

The two lower divisions consist of conglomerates, sandstones and marls, generally of a bright red colour; the Rhaetic alone is marine. In Germany a marine or inland-sea limestone, known as the Muschelkalk, represents the middle part of the Trias and in the eastern Alps the whole series is calcareous and generally dolomitic. The calcareous facies is not represented in Britain, except to a very small extent in the Rhaetic.

Triassic rocks first appear on the south coast a little east of the mouth of the Exe, and form a band extending due north to the shores of the Bristol Channel; they are however in places concealed under Upper Greensand and Chalk. They wrap round the flanks of the Mendip Hills and extend in a gradually widening band up the Severn valley; a little north of Worcester the outcrop suddenly becomes very wide, and covers a large part of Worcestershire, Warwickshire, Leicestershire and Staffordshire. At the southern end of the Pennine Chain the outcrop divides into two, one branch occupying almost the whole of Cheshire and a broad strip on the Lancashire coast. After a short interruption at Morecambe Bay the Trias again appears in Furness, and along the coast of Cumberland as far as Whitehaven. From Maryport it spreads out widely over the plain of north Cumberland and into Dumfriesshire, and also accompanies the Permian outcrop up the Eden valley. In Scotland there are one or two very small scattered patches in Ayrshire and Arran, along the coast of the Moray Firth, and in the Inner Hebrides; Trias is also seen on Belfast Lough and on the coast of Antrim. The eastern branch runs due north from Nottingham, forming the wide plain of the Trent and the Vale of York, passing out to sea at the mouth of the Tees. It will be seen therefore that the Trias forms some of the best agricultural districts in the whole country.

The Bunter beds of Devonshire consist of two members; about 80 feet of coarse conglomerate at the base and 300 feet of coarse textured red sandstone with occasional pebbles above. The pebble beds consist of well-rounded or oval pebbles up to a foot in diameter, mainly of quartzite, grit and Devonian limestone; the pebbles become smaller towards the north while the proportion of sandy matrix increases. The sandstones above also contain occasional pebbles, and the whole series must have been laid down by rapidly moving water, probably in a great river flowing from the north or north-west, with some addition of material from the south-west, since part of the sand grains can be traced to the metamorphic areas around the granites of Devon and Cornwall.

The Keuper beds of Devonshire comprise about 1200 feet of red and green marls, with occasional beds of sandstone, layers of gypsum and casts of cubes of rock-salt. The whole series must have been formed in still water under desert conditions. The uppermost layers for about 100 feet have been bleached by water percolating from above, and are called the Tea-green Marls. Above this comes the Rhaetic marine series, from 30 to 40 feet thick. The latter is well exposed at several points on the shores of the Bristol Channel, but inland it is hardly seen. Around the Mendip Hills the Bunter is absent, and the Keuper is partly composed of wedge-shaped masses of a peculiar breccia, generally called the Dolomitic Conglomerate. This and the accompanying sandstones pass into the more normal Keuper marls.

The following table shows the succession of the Trias beds as seen in Worcestershire. It may be taken as typical of the Midland district:

		Feet
Rhaetic Beds		50
Keuper	{Red marls with rock-salt and gypsum... ...	1000
	{Red sandstone with conglomerate at base ...	300
Bunter	⎧Upper mottled sandstone	200
	⎨Pebble beds	300
	⎩Lower mottled sandstone	200

In general terms it may be stated that the Bunter series of the Midlands consists of coarse sandstones and conglomerates, while the Keuper is composed of fine sandstones and marls, often with rock-salt and gypsum. The Bunter pebble beds thicken towards the north, becoming very conspicuous in Staffordshire and Cheshire, where they form some poor land, as in Cannock Chase. Towards the east the Bunter thins out, and in the north-east of Warwickshire this division is absent. However in Leicestershire and Nottinghamshire the Bunter comes in again and forms among other comparatively poor districts the largely uncultivated area of Sherwood Forest. The pebble beds do not seem to extend beyond Doncaster. In the Vale of York and in the Tees valley the Trias is so largely covered by drift that little is known about it. At Middlesborough a thick bed of rock-salt is worked in the Bunter, all the other British salt-beds being in the Keuper.

In Cheshire the total thickness of the Trias is very great, as follows:

		Feet
Keuper	{Marls and Upper Sandstones	2000
	{Lower Sandstone (Waterstones)	400
Bunter	⎧Upper mottled sandstone ...	600
	⎨Pebble beds	1000
	⎩Lower mottled sandstone ...	400

The pebble beds here are much the same as in Staffordshire but thicker. The Waterstones are so called because they form a good water-bearing horizon. The most important economic product of the Trias is the rock-salt of Cheshire, which forms two beds from 70 to 100 feet thick.

The Trias beds of Lancashire are a continuation of those of Cheshire, the chief difference being that a larger proportion of the sandstones are yellow instead of red. The Trias of Cumberland and the Scottish border is at least 3000 feet thick:

			Feet
Keuper	Red marl and gypsum	...	900
	Kirklinton Sandstone	...	500
Bunter	St Bees Sandstone ·...	...	1500
	Red gypsiferous marl ...	...	300

The most important member is the St Bees Sandstone, of a dull red colour, formerly assigned to the Permian. It is well seen on the coast and in the Eden valley north of Appleby. The softer beds of the Keuper are largely covered by drift.

The outcrop of the Rhaetic beds is everywhere so narrow as to be of no agricultural importance.

As to the fossils of the Trias there is little to be said. It is only in the Rhaetic that a few species of marine invertebrates are found. The Bunter and Keuper, being terrestrial, contain only remains of reptiles, amphibia, fish and plants. The reptilia and amphibia are commonly represented by foot-prints preserved in the fine mud of the lake shores. The most important fossils of the British Rhaetic are the lamellibranchs, *Avicula contorta*, *Protocardia rhaetica* and *Pecten Valoniensis*, with fish, *Acrodus* and *Ceratodus* and reptiles, *Ichthyosaurus* and *Plesiosaurus*. In the Rhaetic beds of the Mendip Hills are found remains of the earliest known mammal, *Microlestes*, a small marsupial.

In the south of Staffordshire the Bunter pebble beds form the greater part of the heathy and forest-clad area known as Cannock Chase. The hills rise with fairly steep sides some 300 or 400 feet above the Trent. Much of the area is open land with heath and sparse oaks or clusters of birch. If enclosed the land is generally under the plough, but a large part of it is hardly worth cultivating, the soils being very light and sandy or gravelly. The Waterstones and Keuper sandstones are thin reddish or white sandstones often rising up as small hills covered with fairly light soil. The Keuper marls are tough red clays with occasional green bands, yielding heavy soils; these form pre-eminently a district of grass land, devoted to dairying, though good crops of corn and roots can be grown. In general terms it may be said that the Bunter forms either land too poor to be cultivated or light arable soils, while the Keuper marl is grass. The Keuper sandstones however resemble the better land of the Bunter.

In Cheshire the pebble beds yield quite a good light soil and carry a fair amount of arable land. The Lower Keuper sandstones generally form high escarpments, as at Helsby and the Peckforton Hills. They yield a light fertile soil, which has been much improved in the past by "marling" with Keuper marl or boulder-clay; this process was very costly, but such land now commands high rents. There is some waste land on this formation in the Wirral peninsula. The Waterstones, a mixture of thin-bedded shales and red and green sandstones of very fine grain, yield rather a heavy soil, mostly in grass. But the leading agricultural characteristic of Cheshire is the prevalence of grass land, which lies on the Keuper marls and on the boulder-clays. Both of these formations yield clay soils which are too stiff and heavy for arable cultivation, but form most excellent permanent pasture, admirably adapted to dairying and especially to the making of cheese, for which this county is noted. In the south of Lancashire the Triassic strata are generally buried under a thick cover of glacial drift of variable character. Here and there the Waterstones come to the surface and form a rather light soil. In north Cumberland and in the Eden valley there is a great thickness of Trias, for the most part sandstones, but also deeply covered by drift and alluvium, which almost wholly determine the character of the soils.

The soils yielded by the British Trias vary greatly in character, including some of the poorest as well as some of the best land in the country. Worst of all are the Bunter pebble beds, which are sometimes uncultivated, heath-covered or forest land. Of the other subdivisions the sandstones tend to form light land, often fertile and well suited to barley and turnips, while the marls are of somewhat heavy character with a preponderance of permanent pasture and dairy-farming.

One of the most fertile districts on the Trias is round Taunton in Somerset, where the soil is a light free-working loam, rather deficient in lime, but admirably adapted for barley; this crop is often taken three times in a five-year rotation. This district lies entirely on the Keuper, which also forms the basement of the great alluvial flats (moors) of central Somerset, in the lower valleys of the Tone and Parret (Sedgemoor and others). This country, which is very like the Fens of eastern England, suffers greatly from floods. In the large Triassic district of the Midlands the soils of the Bunter are generally light and sandy, those of the pebble beds being extremely so; the Keuper sandstones and marls carry some rich soils, often a good deal modified by boulder-clay and drift. In Nottinghamshire the Bunter forms light sandy soils, generally in need of lime, but when well treated they will grow good crops of barley, oats and turnips. Sherwood Forest, on the Bunter pebble beds, is still largely uncultivated. On the Waterstones of the Keuper the prevalent soil is a red, slightly calcareous, clayey loam; a bed of greenish clay at the base forms a stiff yellow soil. The Keuper marls form a stiff red calcareous clay, fertile but difficult to work and needing deep drainage; occasional beds of sandstone, locally called skerries, form loamy and sometimes stony soils. As before stated the soils of the Vale of York can scarcely be regarded as truly Triassic; they are derived almost exclusively from river alluvium and boulder-clay.

CHAPTER XIII

THE JURASSIC SYSTEM

We now come to what is admittedly the most difficult task of the agricultural stratigraphist, namely, the description of the rocks composing the Jurassic system. These rocks show extreme variation of lithological character, both vertically and horizontally. In any given locality the character of the sediment deposited during this period changed very rapidly, so that the succession of rock types is complicated, and furthermore when traced from place to place the rocks composing many of the subdivisions show remarkable changes of facies; consequently there may be little or no resemblance in the soils yielded by the same formation in different localities. For this reason the information afforded by geological maps is, in respect of this system, particularly misleading and must of necessity be supplemented by local knowledge. Hence it is very difficult to give in a reasonable compass a clear account of this most important formation.

The distribution of the Jurassic system, taken as a whole, is particularly simple; it forms a broad band stretching across the country from Dorsetshire to Yorkshire, the only interruption being that due to the overlap of the Cretaceous in the East Riding of Yorkshire, on the western flank of the Wolds. There are two small outliers in Shropshire and Cumberland respectively and a few small and unimportant outcrops in the north-east of Ireland and on both sides of the northern part of Scotland.

It is unfortunate that the nomenclature of this system is in a somewhat confused state. It was originally described as the

Oolitic series, from the prevalence of oolite limestone in the middle members. At a later stage this name was used by some writers for the whole system as at present defined, by others for a part of it only. Hence this name should be avoided. The following table shows the divisions as at present adopted by most authorities:

Upper Jurassic	⎧ Purbeck Series. ⎪ Portland Series. ⎨ Kimeridge Series. ⎪ Corallian Series. ⎩ Oxfordian Series.
Middle Jurassic	⎰ Bathonian or Great Oolite Series. ⎱ Bajocian or Inferior Oolite Series.
Lower Jurassic	Lias.

Over a large part of the country the Portland and Purbeck beds are absent, and the thickness of the different subdivisions varies greatly. In Dorset, where the succession is most complete, the thickness is about 3500 feet, but in the Midlands and Yorkshire it is much less. Nevertheless owing to the low dip the outcrop is very wide even in these districts, especially just south of the Wash. The Jurassic rocks form a region of the most varied relief, ranging from the steep and rugged hills and moorlands of north Yorkshire (Cleveland and Hambleton Hills), the Cotteswolds and the hills of Dorset, to the level plains of the eastern Midlands (Huntingdonshire, Cambridgeshire and south Lincolnshire). Almost the whole of the great plain of the Wash is underlain by Jurassic clays, although they are concealed from view by alluvium and peat. It is perhaps worthy of mention that the comparatively soft rocks of the Lias form some of the highest and steepest cliffs on the English coast. This is the case both in Yorkshire and in Dorset, and is due largely to the fact that the Lias is overlain by much harder rocks, forming the cap of the cliff.

So far as it is possible to generalize on the lithology of the Jurassic system as a whole, it may be said, contrary to the general impression, to be a *clay* formation. It is true that sandstones and limestones are very abundant, appearing in some localities to be the dominant rock-type, but this appearance is in part deceptive, since the hard rocks make a great

show in cliffs, escarpments and quarries, while the softer
members form low ground and are largely concealed by superficial deposits. Limestones are most strongly developed in the
south-western district, clays in the Midlands and sandstones
in Yorkshire.

The general stratigraphy of the Jurassic system has been
studied with great minuteness, perhaps more so than in any
other formation. It is found that the variations of facies and the
occasional gaps in the system are largely due to repeated small
crust-movements, following the lines of axes of folding initiated
at much earlier periods; these are called *posthumous* movements.
These small disturbances were not much in evidence during the
Liassic period, but reached their maximum in the Middle
Jurassic, afterwards again becoming less conspicuous, till at
the close there was a sudden recrudescence, leading in many
places to a strong unconformity as the base of the Cretaceous.

A problem of some interest is the source of the material
composing the Jurassic rocks. The dominant lithological types
are limestones and blue clays and shales, with sands only in the
shallower water. Many of the argillaceous rocks are very
bituminous, some being capable of combustion; this abundance
of carbon is of some significance with regard to their origin,
since it suggests derivation from Coal-measures. The material
for the very abundant limestones might well also be derived
from the Carboniferous Limestone. Much of the material must
have come from land to the north, as indicated by the prevalence
of estuarine conditions in Yorkshire in Middle Jurassic times
and indications of a shore-line in this direction in the Oxfordian
and Corallian. In the south material must have been derived
from the Devonian peninsula and the Mendip Hills, as well as
from the broad ridge of old land that then clearly joined the
Mendip region to the Carboniferous of Belgium, the so-called
London plateau, whose existence has of late years been proved
in so many deep borings. Against this old land Triassic,
Jurassic and Lower Cretaceous strata abutted and overlapped,
till it was finally submerged in the middle of the Cretaceous
period.

I. Lower Jurassic. The Lias

The Lias forms a well-defined series with distinctive characters, differing markedly from the succeeding Middle Jurassic strata. It is also a good deal more uniform, when traced across country, than any of the higher divisions. The thickness also is fairly uniform, from 800 to 1000 feet.

The Lias is essentially a clay formation; in the lower part especially there are a good many calcareous bands, but these usually alternate in thin layers with shale or clay. At certain horizons there are bands of ironstone of great economic value; the chief of these being in the Cleveland district of Yorkshire.

As a matter of practical convenience the Lias is usually divided into Lower, Middle and Upper, but these divisions do not correspond to any clearly marked differences, either lithological or palaeontological. In some districts the demarcation of them is somewhat uncertain. In a more strictly scientific way the subdivisions are effected by means of ammonite-zones, this being the characteristic fossil-group of the whole series. About fourteen principal zones are recognized, with numerous sub-zones. By means of these the succession in distant areas can be correlated.

The Lower Lias as seen on the Dorset coast near Lyme Regis consists of about 500 feet of grey shales, clays and marls, with many bands of limestone, especially near the base. Most of the beds of shale and clay are more or less calcareous, and fossils are abundant; the bands of limestone are often only a foot or two thick, alternating with shale. The origin of the limestone bands is supposed to be mechanical, that is to say, they consist of calcareous deposits formed by denudation of some older limestone, probably the Lower Carboniferous. The Middle Lias is somewhat more sandy, and about 350 feet thick, while the Upper Lias of the south-western district is very thin, certainly not more than 70 feet and possibly much less. Round the Mendip Hills and in Glamorgan the Lias is very thin, being in fact a shore-deposit, the whole series at the same time becoming much more calcareous, and in South Wales there are some massive white limestones. When followed up the Severn valley the series again becomes thicker and more argillaceous. In Warwickshire the Lower Lias is mostly clay, but the upper part of the middle division consists of the well-known Marlstone, a ferruginous and sandy limestone that caps much of the high ground, rising to over 700 feet at Edge Hill. The Upper Lias of this area is mainly clay or shale, and generally about 150 feet thick.

Between Rugby and Lincoln the Lias maintains the same general character, but becomes somewhat thicker; the Lower Lias alone near Grantham is about 700 feet thick, and at the top of it is a valuable bed of ironstone, largely worked in Lincolnshire. The Marlstone is also worked for iron near Grantham. The Upper Lias near Lincoln consists of a series of shales with calcareous concretions and thin limestones, the total thickness being about 100 feet.

In Yorkshire the total thickness of the Lias is about 1100 feet. The Lower Lias is as usual an alternation of shales and thin limestones, about 700 feet thick, but the middle division is much more sandy; moreover it contains some extremely important seams of ironstone. The Cleveland main seam at Eston reaches the great thickness of 25 feet of solid ironstone; besides this there are several other seams, giving rise to one of the largest iron industries in the world.

The Upper Lias consists of about 220 feet of shale; at one time the upper part was largely used for the manufacture of alum, but this is now an extinct industry in this district. The jet trade is also moribund; jet is a kind of bituminized fossil wood, found near the middle of the Upper Lias.

The Liassic outliers of Shropshire, Cumberland, Scotland and Ireland are lithologically very similar to the main outcrop; the total area is small, and that in Cumberland is entirely covered by drift. Those in Scotland and Ireland are of no agricultural interest, but a thick bed of ironstone in the Middle Lias of Raasay will doubtless be of great commercial importance some day.

Since the Lias is in the main a clay-formation it generally forms somewhat low ground. There are however important exceptions. Both in Dorset and in Yorkshire it often forms

the lower parts of high cliffs, where the soft Lias has been
protected from denudation by a capping of some harder rock.
Again the Middle Lias (Marlstone) of the Midlands is a very
hard ferruginous limestone, which, although of no great thickness, forms a prominent escarpment in Warwickshire (Edge
Hill), Northamptonshire and in the south of Lincolnshire, near
Grantham. However the Lower Lias, which is the thickest
division, almost everywhere forms either a low plain or the
floor of deep valleys excavated through the harder beds above.
Most of the patches of higher ground scattered about the
Liassic plain of the Midlands are outliers of hard ironstone or
limestone belonging to the Middle Jurassic series.

The soils yielded by the Lias must on the whole be described
as fertile, and there are few barren or uncultivated tracts on
this formation. The Lias generally forms fairly low ground
and consequently it tends to be covered by accumulations of
alluvium (e.g. Sedgemoor in Somerset) and by drift in the more
northern half of the country. Generally speaking the limestones
and ironstones yield loamy soils, forming good arable land,
while the more purely argillaceous beds are rather heavy and
as a rule are very largely under permanent pasture. The
Lower Lias limestones of the south-west of England form
fairly rich brown loams of free-working character and yielding
good crops of corn and roots. In particular the Marlstone of
the Middle Lias forms some rich land in Dorset and south
Somerset. The clays of the Lower and Middle Lias in this area
are often cold and wet, forming heavy land in the valleys,
mostly under grass, and in places carrying productive orchards
for cider-making. These heavy lands are largely used for
dairy-farming and especially for cheese-making on permanent
pasture. The upper part of the Middle Lias and the Upper
Lias are lighter in character, yielding loams rather than clays,
with more arable land. The well-known fruit-growing district
round Evesham lies largely on Lias. The soil here is heavy and
not naturally rich, but it is subjected to an intensive system of
cultivation and is thus made very productive. In Warwickshire and Northamptonshire clays predominate and carry the
famous pastures of those shires, arable land being rare unless

the Lias is covered by drift. The Marlstone and part of the
Upper Lias are more ferruginous and calcareous, yielding good
red loams in Rutland and south Lincolnshire, and all over this
area the soils of the Lias plains are to a considerable extent
modified by rainwash carried down from the higher ground of
Northampton Sands and other calcareous and iron-bearing
rocks. Thus their texture and composition is often considerably
improved.

From Leicestershire northwards the outcrop of the Lias is
almost always masked by a thick covering of glacial drift or by
some kind of alluvium; hence it becomes more difficult to give
any general account of the soils. Where free from such superficial
accumulations their general character is much the same as in
the driftless area of the south-western counties, with predominating pasture land on the heavy clays and good arable
loams on the limestones and ironstones. In Lincolnshire the
Lias forms strong land, suited to wheat, beans and mangolds,
though much of it is in pasture largely devoted to grazing and
milk-farming. In north Yorkshire the Lias outcrops mostly at
the bottom of deep valleys, with steep sloping sides capped by
the estuarine sandstones of the Bajocian; hence the Lias soils
are much obscured by rainwash and drift. Where free from
such covering they generally form heavy clays, mostly in
permanent pasture. The Lias plain north and west of the
Cleveland-Hambleton escarpment is so thickly covered by drift
that the underlying rock is a negligible factor in determining
the character of the soils.

II. The Middle Jurassic

The truly marine conditions of the Lias appear to have
come to a somewhat abrupt conclusion; everywhere the water
became much shallower, clays and shales giving place to limestones and sandstones; in some places there is a distinct
discontinuity and in a few localities it is possible to prove
considerable denudation of the Lias before the deposition of the
next succeeding series. This is specially noticeable in north-
east Yorkshire, where in some places as much as 100 feet, or

even more, of Lias has been carried away, the hollows in its
surface being filled up by coarse sandstone and impure ironstone;
throughout the counties of Lincoln, Rutland and Northampton
there is also an unconformity at this horizon, and the base of
the upper series is formed by a pebble bed containing rolled
fossils derived from the Lias.

The succession of Middle Jurassic strata, as seen in the south-
western counties, may be generalized as follows:

Bathonian or Great Oolite Series
- Cornbrash.
- Forest Marble and Bradford Clay.
- Great Oolite and Stonesfield Slate.
- Fullers' Earth.

Bajocian or Inferior Oolite Series
- Inferior Oolite.
- Midford Sands.

The above succession generally holds, although there are
local variations and some of the divisions are discontinuous.
The chief difficulty arises from the fact that the basal sands,
though almost everywhere present, are not always at the same
horizon, as indicated by the fossils. However interesting from
the purely scientific standpoint, this is of little practical
importance. In Dorset the Midford sands are from 150 to 180
feet thick; in Somerset they enclose beds of shelly limestone.
These pass up gradually into the limestones of the Inferior
Oolite proper, which are only about 10 feet thick on the coast,
but near Yeovil they increase to about 50 feet. The lowest
member of the Bathonian in Dorset is the Fullers' Earth,
which attains a thickness of 400 feet; it is a soft marly clay,
in part capable of being used for "fulling" cloth (i.e. removing
grease) and for clearing oils. Next comes the Forest Marble,
a series of shelly and flaggy limestones about 100 feet thick.
This is succeeded normally by the Cornbrash, whose general
characters will be described later.

Perhaps the most typical development of the calcareous
facies of the Middle Jurassic rocks is that seen in the well-known
agricultural district of the Cotteswold Hills. The general
succession is as follows:

		Feet
	Cornbrash	15
	Forest Marble	60–100
Great Oolite Group	Great Oolite	40–100
	Stonesfield Slate	10– 15
	Fullers' Earth	40– 80
	Ragstones	20– 40
Inferior Oolite Group	Freestones and Pea Grit ...	35–200
	Cotteswold Sands	40–120

The Cotteswold sands, though closely resembling the
Midford sands, clearly belong to a higher stratigraphical horizon.
They are thickest just north of the Mendip Hills and thin out
towards the north-east; most of the limestones, on the other
hand, are thicker in this direction. Although the limestones
have been divided by specialists into a great number of minor
groups, such elaboration is unnecessary for the present purpose.
The whole Middle Jurassic succession of the Cotteswold region
above the sands may be briefly and generally described as
consisting mainly of limestones, often markedly oolitic or
shelly, often flaggy in structure, and with occasional intercalations of shale or clay, such for example as the so-called
Bradford Clay at the base of the Forest Marble series; this
band, though well known and conspicuous, is only 10 feet thick.
The limestones vary a good deal in character, and show every
gradation from one type to another. The more compact beds
form excellent building stone, as for example the famous Bath
stone.

When followed into Oxfordshire and Northamptonshire the
Bajocian series becomes entirely sandy and ferruginous, and
partly of estuarine origin, but the Bathonian succession is very
similar to that seen in Gloucestershire. At the base of the
whole succession are some valuable beds of ironstone, extensively worked in Northamptonshire and Rutland. When
followed towards the north the estuarine facies becomes more
and more dominant, the only important marine bed being
the Lincolnshire Limestone; this forms a valuable building-
stone, well-known quarries being those at Clipsham, Ketton and
Ancaster. This series forms the remarkable long, straight
escarpment that extends throughout Lincolnshire in a due north

and south direction. The city of Lincoln is situated at the point where it is cut through by the river Witham. At the base of the Lincolnshire Limestone in the south of the county is a layer of thin-bedded sandy limestone much used for roofing purposes, the so-called Collyweston slate; this is of course flagstone, and not true slate. The Bathonian series of Lincolnshire consists partly of clays of various colours, the representative of the Great Oolite limestone being only about 20 feet thick. The Cornbrash at the top is as usual (see below). The so-called "heath" land of Lincolnshire lies on these formations. The soils are light and well adapted to the growth of barley, roots and seeds, also the rearing and fattening of sheep.

In the East Riding of Yorkshire the Middle Jurassic is for the most part concealed by the overlap of the Cretaceous, but in the North Riding there is an extensive development of beds of this age, for the most part estuarine in origin, with a total thickness of about 650 feet. There are one or two thin calcareous beds representing the limestones of the Midlands and the succession is terminated by the Cornbrash. The estuarine beds of Yorkshire are a very variable succession of sandstones and shales, with plant remains and occasional thin seams of coal, the whole showing a strong resemblance to the Coal-measures on a small scale. At the base is the Dogger, which is in some places a ferruginous sandstone, but in other places a valuable seam of ironstone. The whole series is well seen in the cliffs between Scarborough and Saltburn, and forms the capping of the Cleveland and Hambleton Hills, which rise to a height of over 1400 feet. A great part of the outcrop is open moorland, intersected by deep valleys cut through into the Lias.

The Cornbrash is one of the most curiously persistent beds in the whole Jurassic succession, hardly varying in thickness or in character from one end to the other of its outcrop. It consists of light-coloured earthy or rubbly limestone, often rich in fossils and generally from 5 to 15 feet in thickness, though locally it may be somewhat thicker. The limestone is said to be unusually rich in phosphoric acid, at any rate in the south of England, and this may partly account for the fertility of the soil yielded by it.   In the north it is more ferruginous and sometimes oolitic, the soils being less fertile.

One of the most noticeable features of the calcareous facies of the Middle Jurassic strata is the abundance of oolitic and pisolitic limestones, from the prevalence of which the whole formation was formerly known as the Oolitic system.   Limestones of this type are formed in shallow but clear water, often in the near neighbourhood of coral reefs, where calcareous organisms are abundant and the water easily becomes comparatively rich in calcium carbonate.   The mode of origin of such rocks has been discussed in an earlier chapter (see p. 91). Many of the beds of oolite form excellent building stone.   The beds of clay indicate either a temporary deepening of the water or an influx of mud from some distant source.   The estuarine strata generally show a rapid alternation of sandstones and shales, and present every indication of having been formed as delta deposits at the mouths of large rivers.   Occasional thin marine bands show temporary incursions of the sea, probably due to slight local depression.   The ironstones are almost invariably formed by metasomatic replacement of limestones as described on p. 97.

The fossils of the Middle Jurassic strata naturally vary in character according to the conditions of formation of the sediment.   The limestones and other marine beds contain a remarkably abundant marine fauna, specially rich in ammonites. Lamellibranchs, gastropods, brachiopods and corals are also very abundant, and many of the limestones consist almost exclusively of more or less broken shells.   It has been found possible to divide the marine facies into zones by means of their characteristic fossils, chiefly ammonites, thus:

Bathonian	Cornbrash	*Amm. macrocephalus.*
	Forest Marble	*Waldheimia digona.*
	Great Oolite	*Nerinaea Voltzi.*
	Fullers' Earth	*Amm. subcontractus.*
Bajocian	Upper Limestones	*Amm. Parkinsoni.* *Amm. Humphriesianus.*
	Lower Limestones	*Amm. Murchisonae.*
	Midford Sands	*Amm. opalinus.* *Amm. jurensis.*

This table applies to the south-western succession; the correlation of the estuarine facies presents some difficulties, but it seems clear that the Lincolnshire Limestone belongs to the zone of *Amm. Murchisonae*, while the Scarborough lime-stone, the chief calcareous bed in Yorkshire, belongs to the *Humphriesianus* zone. All the beds between this and the Cornbrash are estuarine.

Owing to the varying lithological character of the beds composing the Middle Jurassic series they offer unequal resistance to denudation, and as a consequence the topography of the area occupied by these strata shows much variation. In the south-west of England the limestone bands are the hardest and form strongly marked escarpments, with the steep face towards the north-west and a gentle dip-slope to the south-east; this for example is the general structure of the Cotteswold Hills. The Bath Oolite and the Forest Marble also form con-spicuous escarpments in Somerset and Dorset, often rising into flat-topped hills, whereas those composed of Midford Sands are commonly conical in shape. The Fullers' Earth and other beds of clay form valleys between the escarpments. When followed through the Midland counties the Bajocian and Bathonian limestones and sandstones usually constitute fairly high ground, often in distinct escarpments, strongly contrasting with the plains of the Lias on the one hand and of the Oxford Clay on the other. As already mentioned there are scattered over the Midland counties many outliers of these rocks resting on Lias. Owing to differential erosion and faulting the distribution of this series is almost everywhere very complex at the surface, and consequently the soils show great variety within short distances.

Owing to the rapid variations, both vertical and horizontal, in the character and composition of the Middle Jurassic rocks, it is almost impossible to give any general account of the soils yielded by them. Furthermore over the northern half of their outcrop, from Northamptonshire to Yorkshire, the matter is complicated by a widespread cover of glacial drift of varying character. In general terms it may be said that the clays and shales yield heavy soils, difficult to cultivate and largely under permanent pasture, devoted to stock-feeding and dairy-farming; the limestones on the other hand carry light loamy soils, often rather stony, and eminently adapted to barley and turnips; they are rarely strong enough to give the heaviest yields of wheat, although this crop is extensively grown. Such soils are liable to suffer from drought in dry seasons, and water supply is often a difficulty. The sandstones of the Midlands often give rise to light soils of good quality, but in the moorland district of Yorkshire they are largely uncultivated and covered with heather and peat; owing to the high elevation they are devoted to Blackface sheep. The character of the soils lying on some of the more distinctive subdivisions may be very briefly mentioned. The Midford and Cotteswold sands carry light soils but are limited in their extent. The limestones of the Inferior Oolite form what are often called *brashy* soils, that is, fairly light loams, often rather stony and well adapted to corn growing, especially barley, and also suitable for turnips. These limestones form the soils of the well-known Cotteswold sheep district. The Fullers' Earth on the other hand is very heavy and wet, and is mostly under grass. The Great Oolite limestone forms a rather thin "brashy" soil, suitable for turnips and barley, but the clay part of this division is a heavy soil. The Cornbrash as its name implies is a good corn-growing soil, but in the northern part of its outcrop it is not so fertile as in the south-west of England. The Lincolnshire Limestone forms a light and not very productive soil, while the estuarine clays of the region between Northamptonshire and the Humber are rather stiff and of poor quality. In the south of Yorkshire the outcrop of this series is very narrow; further north it expands again and forms the large area of high-lying ground known as the North Yorkshire moors, whose character has already been sufficiently indicated.

III.　The Upper Jurassic

The rocks composing the Upper Jurassic series show less variation than those of the Middle Jurassic, but even here most of the major subdivisions comprise at least two distinct facies. The distribution of these rock-types is however easily explained

on the assumption that the British area then consisted of a
sea basin, deep over the Midland counties, shallower towards
the south-west and the north-east, with in all probability a
shore-line at no great distance in both directions.   In the later
stages land encroached on the sea in the south of England.
while in the north the sea became deeper.

The generalized classification of the Upper Jurassic, as
usually adopted, together with the prevailing character of the
rocks, is shown in the following table:

	South of England	Midlands	North of England
Purbeckian ⎫ Portlandian ⎰	Limestone	Absent	Absent ?
Kimeridgian ...	Clay	Clay	Clay.
Corallian ...	Limestone	Clay	Limestone.
Oxfordian ...	Clay	Clay	Sandstone.

Hence it appears that this is largely an argillaceous
formation, the limestones and sandstones being local and
subordinate.

The Oxfordian subdivision takes its name from the Oxford
Clay, which in the stretch of country between Oxford and the
Wash forms practically its whole thickness.   However both in
the south-west of England and in Yorkshire the lower part consists of sandstones, often with a calcareous cement, known as
the Kellaways Rock; this is a shallow-water facies of the
lower zones of the Oxfordian, indicating approach to shore-lines
in both directions.   In Dorset the Kellaways beds are about
70 feet thick, but they become thinner to the north-east and
disappear in Bedfordshire.   In north Yorkshire the Kellaways
rock sets in again and attains in the moorland district a thickness
of about 100 feet, forming some very bold features.   The
Oxford Clay, which forms the surface rock of a very large area
in the eastern Midlands, is a stiff, bluish, greenish or grey clay,
becoming brown or yellow at the surface owing to oxidation.
Sometimes it is rather shaly, sometimes quite without signs of
bedding, breaking only with a conchoidal fracture into irregular
lumps.   There are occasional harder calcareous bands and lines
of concretions, while in some places large crystals of gypsum

(selenite) are abundant. Fossils are numerous, especially ammonites, belemnites and large oysters (*Gryphaea dilatata* and allied forms). The ammonites are usually preserved in iron pyrites.

The Oxfordian may be conveniently divided by means of its ammonites into three zones:

Zone of *Ammonites cordatus*,

Zone of *Ammonites ornatus*,

Zone of *Ammonites calloviensis*.

These are sometimes further divided into sub-zones. *Amm. calloviensis* seems to be confined to the sandy Kellaways facies; and is scarcely known where the whole series is argillaceous, as in Huntingdonshire. In some localities remains of reptiles are very abundant.

Owing to the great uniformity of the Oxford Clay it is unnecessary to describe its distribution in detail; from the Dorset coast to the Humber it varies only in the extent to which it is replaced by the Kellaways facies; in North Yorkshire more than half the formation is Kellaways rock, and the rest is a grey sandy shale with few fossils. This however mostly forms uncultivated moorland, and is of little importance.

In the Corallian series the existence of two distinct facies is very conspicuous; both in Dorset and in Yorkshire this division consists of an alternation of calcareous sandstones and limestones, generally oolitic; in the Midland counties these are replaced by a clay, very similar to the Oxford Clay and often confused with it, as for example on the published maps of the Geological Survey. Owing to the great difference in the hardness of the two rock types the topography of the Corallian country shows strong variations. In Dorset and Wiltshire and in Yorkshire it forms elevated ground with conspicuous escarpments, while in the Midlands it forms part of the great plain of the Wash drainage basin.

The Corallian series of the north of England, as developed in the neighbourhood of Pickering, shows a rather complicated series of sandstones and limestones with an occasional shaly band:

	Feet
Upper Calcareous Grit	40
Upper Limestone and Coral Rag } Coralline Oolite ... ... }	50
Middle Calcareous Grit	80
Lower Limestones	100
Lower Calcareous Grit	130
	400

In the above table the succession is simplified as much as possible and the thicknesses given are the maximum for each division. The Calcareous Grits are massive yellowish sandstones with a calcareous cement, often weathered into striking tabular and columnar forms. The Lower Calcareous Grit forms a very well-marked escarpment facing north, and caps many of the tabular hills of the moorland district.

The limestones vary a good deal in character; some beds are oolitic or pisolitic, while others are very rich in corals and shells. Part of the lower limestone series is to a certain extent silicified and converted into chert, especially near Filey. The term Coral Rag is commonly used to describe the massive shelly non-oolitic varieties of limestone, some of which appear to be true coral reefs.

In Dorset and Wiltshire the general succession is similar; the Lower Calcareous Grit appears to comprise also the representatives of the Lower Limestones and the Middle Grit: that is, the sandy facies extends higher in the series. The upper limestones are very similar to those in Yorkshire, but near Westbury they are largely converted into a valuable bed of iron ore. A few miles east of Oxford the limestones come to an end and from Buckinghamshire to the Humber the clay facies alone is found, except for a small coral reef at Upware, near Cambridge.

The argillaceous representative of the Corallian, the Ampthill Clay, was laid down in fairly deep but muddy water and consists of a dark bluish grey clay with, in some places, a band of ferruginous oolitic limestone at the base. The total thickness is somewhat uncertain; it probably does not anywhere exceed 50 feet, and may be much less. Most of the outcrop of the Ampthill Clay is covered by drift and alluvium. It is almost invariably mapped as Oxford Clay, but can be distinguished in the field from this formation by the fact that the fossils are preserved in calcite instead of iron pyrites. Agriculturally it is almost exactly similar to the Oxford Clay and the line of demarcation is uncertain, both above and below.

There can be no doubt as to the geographical conditions prevailing in Britain in Corallian times. The limestones and sandstones of the north and south-west were laid down near to shore-lines, in clear water with strong currents and abundant animal life; a warm climate favoured the growth of reef-building corals. In the central part of the basin the sea was deeper but muddy, and therefore unfavourable for calcareous organisms. The mud was probably derived from the denudation of an ancient land now hidden under the Cretaceous rocks of south-eastern England.

The Kimeridgian series is a much more uniform formation than the Corallian; it consists entirely of argillaceous strata, the Kimeridge Clay; this, though varying considerably in thickness in different parts of the country, shows very constant lithological characters. It consists of dark grey or black clay, usually more or less shaly, with bands of calcareous nodules and crystals of selenite. Occasionally there are to be seen thin layers of grey or whitish impure limestone or hardened marl. In some places the clay is very bituminous and fossils are commonly abundant, though as a rule not well preserved.

The Kimeridge Clay on the Dorset coast is about 1000 feet thick, but it decreases rapidly towards the north-east; from Oxfordshire to the Wash there is only about 100 feet; this is partly to be accounted for by denudation before the formation of the overlying Cretaceous rocks, but in the main it is due to a real diminution in the original thickness on passing towards the centre of the basin of deposition. In Lincolnshire the thickness increases again to about 600 feet; in Yorkshire, owing to concealment by drift and alluvium, the thickness is uncertain, but is probably not less than in Lincolnshire.

Owing to its softness the Kimeridge Clay nearly always forms low ground. In the south-west of England the surface exposures are not large, owing to the overlap of the Cretaceous, and in the central portion the outcrop is narrow on account of the small thickness of the clay itself. The Kimeridge Clay certainly underlies a large part of the Fenland and in Lincolnshire it forms the eastern half of the broad plain between the Cliff and the Wolds. In Yorkshire it forms the basement of at any rate the greater part of the Vale of Pickering, although there are no exposures and the clay is deeply buried in drift and alluvium. It can however be seen on the coast near Speeton.

Rocks of the Portland and Purbeck series cover only a very small surface area in the British Isles. Where best developed on the Dorset coast the Pórtlandian series comprises two subdivisions, the Portland Sand below and the Portland Stone above. The sandy beds vary in thickness from 40 to 160 feet but are of little interest or importance. The Portland Stone series has a total thickness of about 100 feet and includes the famous building-stone, together with some silicified cherty beds. The building-stone is a very white oolitic limestone, passing into a white compact rock very like Chalk, even containing flint or chert in nodules. The whole group thins towards the north; in the Vale of Wardour, west of Salisbury, it is only about 100 feet thick, and at Swindon and near Oxford almost the whole of the series is sandy. In Buckinghamshire the lower part of the Portlandian is represented by the Hartwell Clay, with sands and thin limestones above.

Rocks of the Purbeckian series are only known with certainty to come to the surface in the Isle of Portland, the Isle of Purbeck, the Vale of Wardour, and a small patch of undefined extent in the middle of the Weald. The latter is usually omitted from geological maps. In Dorset and Wiltshire they consist of thin limestones, shales and marls, formed under varying conditions, marine, estuarine and fresh-water; some black earthy layers are actually old surface soils, with stumps of trees still rooted in them. Many of the beds are very rich in fossils, and the shell-limestones are quarried under the name of Purbeck Marble. The total thickness is about 400 feet. In the Vale of Wardour the thickness is only about 100 feet, and the area exposed is small. Near Swindon

and in Buckinghamshire are some thin beds with fresh-water
fossils that may be of Purbeck age.

Portland and Purbeck beds are not again seen south of the
Wash, and it is still uncertain whether they are represented by
any part of the Spilsby Sandstone in Lincolnshire. In Yorkshire the uppermost Jurassic and Lower Cretaceous strata are
represented by the Speeton Clay, a thin black clay very like
the Kimeridge, only exposed on the coast. This may underlie
the southern part of the Vale of Pickering, as it is said to have
been seen in borings at the foot of the Wolds.

The Oxford Clay forms some of the heaviest soils in the whole
of the British Isles. They are difficult and expensive to work,
and owing to the flatness of most of the ground, nearly always
in need of draining. In Huntingdonshire and the Fenland,
where the outcrop is widest, there is generally a thick covering
of drift and alluvium, and the true surface of the Oxford Clay
is often below sea-level. The soils of the Corallian division
naturally show much variation; the Calcareous Grits of Yorkshire form high ground bearing a light sandy soil, much of which
is moorland. The limestones which form the southern part of
the hilly region stretching for some 40 miles west of Scarborough,
and also the western part of the range bounding the Vale of
Pickering on the south yield on the contrary good loamy soils,
often rather light and stony but carrying good crops of corn.
Where this rock-type reappears in the south-west of England
from Oxford to Dorset the soils are again of very similar character. The Ampthill Clay of the Midlands is almost exactly
like the Oxford Clay, but is mostly covered by drift. The
Kimeridge Clay forms almost uniformly a heavy rather cold
soil, covered by drift and alluvium in the Fenland, Lincolnshire
and the Vale of Pickering. Further south it forms the soil of
the important dairy-farming district of the Vale of Aylesbury
and of the Vale of White Horse, and reappears in some force in
the Vale of Wardour in Wiltshire. The clays of the Upper
Jurassic, where free from drift, form low-lying damp districts
which are mostly under permanent pasture and specially
suitable for dairying.

CHAPTER XIV

THE CRETACEOUS SYSTEM

Towards the end of Jurassic times a series of earth-movements began which produced important results in English stratigraphy. In the south of England the estuarine and fresh-water type of deposit continued for some time; in Yorkshire on the other hand the sea remained deep and in it was formed the Speeton Clay, which may be regarded as a continuation of the Kimeridge Clay. Over the tract of country from Oxfordshire to the Wash the disturbance was most strongly felt and produced a well-marked unconformity, accompanied by considerable denudation of the deposits just formed. In some part or other of this tract the base of the Cretaceous rests on each of the subdivisions of the Jurassic from the Oxford Clay upwards. Between the Wash and the Humber there is not much indication of unconformity, but it is very doubtful whether the succession is complete. Not only are the higher beds of the Jurassic absent in the eastern Midlands, but the lower divisions of the Cretaceous were never deposited. There was a broad ridge of land, with its highest part in Bedfordshire or Cambridgeshire, separating two seas; this ridge was gradually submerged during the earlier part of Cretaceous time and the successive marine deposits overlapped one another from each side towards the centre; by the beginning of Upper Cretaceous time the ridge was completely submerged, as well as the old land of the London plateau, which was undoubtedly the source of much of the material of the Lower Cretaceous strata.

The Cretaceous system falls naturally into two divisions, Lower and Upper; owing to the great variability in thickness

and character of the Lower Cretaceous it is difficult to devise
a classification and nomenclature of general application.   The
names usually applied are taken from the development as seen
on the coast of Kent and Sussex, but they are strictly of local
application only, being unsuited to the rocks as seen in Devon,
Yorkshire or Ireland.   These facts can be expressed in the
general statement that the Cretaceous system shows much
variation of facies, and that the southern facies is usually
accepted as typical.   On this basis the classification is as follows:

| Upper Cretaceous | { Chalk. Gault. |
| Lower Cretaceous | { Lower Greensand. Wealden. |

The local distribution and variations of facies in each of
these divisions will be described separately.

The Wealden series. This series, which is sometimes
called the Neocomian, is somewhat limited in its distribution
in Britain.   The estuarine and fresh-water facies, the Wealden
proper, is found only south of the Thames.   Over the area
between the Thames and the Wash it is absent, but comes in
again in Lincolnshire and Yorkshire as a marine facies.   It is
only in the Weald district itself that this formation covers any
considerable area.

In Kent, Surrey and Sussex the Wealden series is divided
on lithological grounds into two parts, the Hastings Sands below
and the Weald Clay above.   The topography and soils that
prevail on these two formations differ a good deal.   The
Hastings Sands form the central part of the Wealden district,
rising in Crowborough Beacon to a height of nearly 800 feet.
The maximum thickness is about 1000 feet and three divisions
are recognized; the Wadhurst Clay, in the middle, about 150
feet thick, separating two main masses of sand and sandstone, the
Ashdown Sand below and the Tunbridge Sand above.   The soils
yielded by this formation are somewhat variable in character,
but mostly poor.   Some of the sands are so fine in texture that
in practice they are almost as heavy as clays.   They are mottled
yellow and green a little below the surface and there is a strong
tendency to the formation of ironstone concretions.   The soils

are always markedly deficient in carbonate of lime. A large proportion of the High Weald country on the Hastings Sands is wooded and there is a considerable area of unenclosed land in Ashdown Forest. The lower slopes are mostly rather poor grass and it is only on the floors of the valleys in the eastern part of the area that any good farming is found. As a rule neither turnips nor barley are grown. In the better districts hops are cultivated to a certain extent. The chief industry is stock-farming on grass land, with dairying near the railway lines.

The Weald Clay has a maximum thickness of about 900 feet in the west, thinning to the east. It consists of blue and brown clays with occasional layers of shelly limestone and sand. The fossils are chiefly fresh-water shells. In strong contrast to the sandy strata above and below it, the Weald Clay forms a flat plain which is a striking feature in a contoured map of the district. This is an area of extraordinarily heavy soil, and is now nearly all grass and woodland. There is very little arable land and on this neither turnips nor barley are grown, though wheat, beans and mangolds do fairly well. The grass land is poor in quality and is chiefly devoted to the breeding of Romney Marsh sheep and Sussex cattle. The small acreage under hops is rapidly decreasing.

The Weald Clay forms a stiff greasy clay with nodules and concretions of iron oxide in the subsoil. These are sometimes almost numerous enough to form a pan, and to require breaking up with a crowbar before trees can be planted. Some samples of these soils contain as little as 10 per cent. of sand, the rest being fine silt and clay. The most notable chemical characteristic of all Weald soils is the deficiency in calcium carbonate.

The Weald Clay is also exposed to a small extent in the Isle of Wight and in Dorset, near Swanage, where it is over 2000 feet thick but the base is not seen. It consists of variously coloured clays with occasional sand beds; the upper part is chiefly grey shale. Owing to the steep dip the outcrop in Dorset is narrow. In the Vale of Wardour in Wiltshire the Wealden series is reduced to about 100 feet of clay and beyond this no beds of this age are seen south of the Wash. In Lincolnshire the Wealden series appears to be represented by the whole or part of the Spilsby Sandstone; possibly also the Claxby iron-stone and Tealby Clay may belong here, but the correlation of the Lower Cretaceous rocks of Norfolk and Lincolnshire is so uncertain that nothing definite can be said on this point. In Yorkshire the equivalent is to be found in the lower and middle zones of the Speeton Clay. In neither case is the area of the outcrop large enough to render the question of any agricultural importance.

The Lower Greensand series. Where most complete, in the south-eastern counties, this series consists of a bed of clay at the base, the Atherfield Clay, followed by a considerable thickness of sand and sandstone. The whole series is marine, in contrast to the fresh-water and estuarine deposits of the Wealden. In the Isle of Wight the total thickness is about 800 feet, but towards the north it becomes much thinner owing to the gradual dying out of the lower divisions; in parts of Cambridgeshire it appears to be absent altogether. In west Norfolk and in Lincolnshire lower beds again come in, but in Yorkshire the Speeton Clay facies of deposition still continues. The Lower Greensand between Dorset and the Wash affords an excellent example of unconformable overlap, the successive beds gradually extending further and further from either side over a ridge of Jurassic rocks whose summit lay on the borders of Bedfordshire and Cambridgeshire. It was not until Upper Cretaceous times that this ridge was completely submerged. Hence it follows that south of the Humber the Lower Greensand consists entirely of shallow-water deposits, for the most part coarse sands with occasional pebble beds. The name Greensand refers to the occurrence in many places of a conspicuous amount of glauconite, imparting a green colour to the unweathered sands; this green colour disappears when the sand is at or near the surface of the ground, owing to oxidation of the iron. Agri-culturally the Lower Greensand is important as a soil former in two areas, namely, (a) Kent, Surrey and Sussex; (b) along the outcrop from Dorset to the Humber. Each of these will be dealt with separately.

The south-eastern area. The Lower Greensand forms a

strip of generally elevated country, of varying width, all round
the Weald, except on the coast. On the north the outcrop runs
westwards from Folkestone to Farnham; then it turns due south
to Midhurst, whence it runs east-south-east at the foot of the
South Downs to Eastbourne. Leith Hill, 965 feet, chiefly
composed of Lower Greensand, is the highest point in the
south-east of England, and the well-known high ground about
Hindhead and Haslemere is also on this formation. The greatest
width of the outcrop, near Godalming, is about 12 miles, but
the southern belt is much narrower.

In this district four subdivisions are recognized, namely:

Folkestone beds	Green and grey sand and sandstone with bands of chert	80 feet.
Sandgate beds	Sand, clay and Fullers' Earth	70 ,,
Hythe beds	Greenish yellow sand and hard grey calcareous sandstone (Kentish Rag)	200 ,,
Atherfield Clay	Blue clay with marine fossils	60 ,,

The Atherfield Clay is a stiff blue clay, weathering brown,
very like the Weald Clay, but distinguished from it by the
presence of marine fossils. Though naturally very heavy
the soil is modified by downwash of sand from the beds above,
which form higher ground; hence the soil is a fertile loam.

The Hythe beds carry very different soils in different parts.
In Kent the greater part is a free-working, though rather stony
loam, admirably adapted to hops and fruit. Around Sevenoaks
hops of the best quality are largely grown. Some argillaceous
beds give a rather heavy soil in places. The hard Kentish Rag
forms a good building stone, though it cannot be dressed
smooth. West of Redhill the soils become almost pure sand
and are almost entirely woodland or open uncultivated commons
covered with heather, gorse, birch and pine. This is now a
very favourite residential district, owing to its high elevation,
dry soil and beautiful scenery. The southern outcrop, though
narrow, carries some good barley soils near Midhurst.

The Sandgate beds are of finer texture and form a rather
heavier soil, which is partly in pasture in the valleys. Some
thin beds of impervious clay cause the soils here to be rather

wet. Beds of Fullers' Earth are worked at Nutfield. Near
Godalming a bed of calcareous rock, the Bargate stone, forms a
light free-working loam, which is well adapted for arable sheep-
farming. The Rother valley in west Sussex is an area of fertile
arable land, also largely under sheep.

The Folkestone Sand is much coarser in texture than the
other divisions, giving rise to very poor sandy soils, largely
uncultivated and forming heaths and commons, sometimes
covered by sour black peaty soil where a pan has been formed.
In recent times a good deal of land has been planted with
Scotch fir and Austrian pine.

To sum up, it may be said that the great majority of the
Lower Greensand soils are very light, and some of them are
almost pure sand and therefore useless for cultivation. Only in
the lowest division, the Atherfield Clay, is there any notable
amount of heavy land. The most strongly marked peculiarity
in all the soils is the almost complete absence of lime, and
humus is generally very deficient. Only in a few low-lying
areas is there any really good soil, doubtless largely due to an
admixture of rainwash and alluvium.

The Lower Greensand series occupies a considerable area in
the southern part of the Isle of Wight and here reaches its
maximum thickness, about 800 feet. Above the Atherfield
Clay, some 80 feet thick, are ferruginous sands and sandstones,
with a good many marine fossils. The uppermost part, locally
known as the Carstone, appears from recent researches to belong
really to the Gault series, and is therefore on a higher horizon
than the Carstone of Norfolk and Lincolnshire.

Western and northern outcrops. The Lower Greensand is seen
in many places along the Cretaceous outcrop from the Dorset
coast to the Wash, but it is not by any means continuous, being
sometimes absent altogether or overlapped by the Gault.
There is a narrow strip of it in Wiltshire, from near Devizes to
Faringdon, and a patch south of Oxford, where it is mainly
occupied by Nuneham Park; a larger and more important strip
extends from Leighton Buzzard to the neighbourhood of
Cambridge; including the well-known areas of Woburn Sands,
Sandy and Potton in Bedfordshire. Beyond Cambridge the

outcrop becomes narrow and discontinuous, forming the caps
of some of the higher lands in the Fens, especially the Isle of
Ely. Even here however it is largely concealed by drift. In
Norfolk again there is an interesting outcrop of this series
extending from Downham Market to Hunstanton. In Lincolnshire this series outcrops over a considerable area on the south
and west of the Chalk Wolds, but in Yorkshire it is represented
only by the uppermost part of the Speeton Clay. The Lower
Greensand in Bedfordshire forms a considerable area of very
light sandy soils, especially around Woburn Sands, Ampthill,
Sandy and Potton. A very large part of this is in parks and
woodlands; the rainwash from the Lower Greensand escarpment when mixed with river alluvium forms a soil eminently
suitable for vegetable growing, as in the Ivel valley near Sandy
and Biggleswade; from here the London markets are largely
supplied. Somewhat further east the outcrop of the Lower
Greensand forms soil suitable for fruit-growing, especially at
Histon and Cottenham, north-east of Cambridge.

The small patches capping the Isle of Ely and other slightly
elevated areas in the Fens are often in their turn covered by
glacial gravels and sands, the soils differing considerably from
the black soils of the lower levels. From King's Lynn to Hunstanton is a strip of sandy land, with much heather and bracken,
largely woodland, but carrying some good light soils, which
however are often much modified by boulder-clay and other
glacial deposits. This district is admirably adapted for game
preserving, as at Sandringham.

The Gault series. The strata lying between the Lower
Greensand and the Chalk show much variation in character in
different parts of the country. As seen on the coast of Kent,
at Folkestone, almost the whole series consists of pale grey clay[1]
with many fossils. Further west, in Surrey and Hampshire,
the upper part consists of sands and sandstones, often containing
glauconite. This sandy facies, the so-called Upper Greensand,

[1] Gault is a Cambridgeshire word signifying originally any stiff grey clay,
including the Oxford, Ampthill and parts of the Kimeridge Clays, as well as
the true Gault. It is still used in this sense by well sinkers and others, whence
has arisen much confusion.

increases in thickness towards the west, at the expense of the clay and in Dorset it comprises almost the whole formation, only a few feet of clay remaining at the base near Lyme Regis. In Devonshire and Somerset the whole is Greensand. When the outcrop is followed from Somerset towards the north-east a similar change is seen in reverse order, the clay becoming thicker and thicker till the Greensand dies out altogether in Bedfordshire. Between this point and the borders of Norfolk the whole series is clay, thinning towards the north; near King's Lynn a further change occurs and at Hunstanton the whole is represented by 3 or 4 feet of a curious red limestone. This reappears beyond the Wash and continues through Lincolnshire to the Yorkshire coast, being thicker than in Norfolk and paler in colour; north of the Wash it is called the Red Chalk.

The best exposure of the Gault is at Folkestone, where it is about 100 feet thick, but it thickens towards the west. It is a stiff bluish or grey clay with occasional seams of glauconitic sand and many phosphatic nodules. Fossils are very abundant and preserved in a peculiar manner, often retaining the pearly lustre and iridescence of the shell.

The Gault clay forms a narrow valley from one to two miles wide all round the Wealden district between the Lower Greensand and the Chalk, both of which form high ground. This valley is due to the more rapid denudation of the soft clay, which is often very wet and full of springs, since it throws out the water from the Chalk above. The Gault weathers into a brown or nearly white soil; the latter strongly resembling the Chalk Marl soils above. In places where Gault has been naturally mixed with sand from the higher ground the soil is sometimes good, but the unmodified clay is too heavy for arable land and is nearly all in pasture. The total area is not large, and in the west where the outcrop is widest it is to a considerable extent under oak forest (Alder and Alice Holt, west of Farnham in Hampshire).

In the Isle of Wight the Gault clay is seen chiefly in the cliffs and is known as "blue slipper," from its tendency to produce landslips, bringing down masses of overlying strata and forming the well-known "under-cliffs" of the island.

XIV]   THE CRETACEOUS SYSTEM   277

The Upper Greensand, the sandy facies of the Gault, hardly exists in Kent, the Gault there passing gradually into the overlying Chalk Marl, but in eastern Surrey it begins to assume some importance. In western Surrey and on the borders of Hampshire, and also along the southern outcrop at the foot of the South Downs it forms a strip or terrace of land about a mile wide, somewhat elevated above the low Gault valley. In Surrey and Sussex the Upper Greensand generally consists of about 20 feet of white or greenish-grey calcareous sandstone, sometimes almost soft and incoherent, but in other places forming a good building stone, often known as Malmstone. It thickens towards the west and covers a considerable area round Farnham. All round the Wealden area the Upper Greensand forms a rather heavy soil, dark in colour when wet, but drying very white. It is generally fertile and in some parts carries good crops of hops, wheat and mangolds. There is little wood and no waste land. Near Reigate a thin bed of hard stone is cemented by silica instead of calcium carbonate. This resists heat well, and is much used for hearth-stones.

The Upper Greensand appears again in the Isle of Wight, and in Dorset it reaches its maximum thickness of about 180 feet, thus almost wholly replacing the Gault clay. Isolated patches of the sandstone and chert extend even to the Haldon Hills beyond Exeter. In Somerset the Blackdown beds, which are of about the same age, though perhaps representing a little of the Chalk as well, are of rather peculiar character. The Blackdown Greensand is a highly siliceous rock, even the fossils consisting of silica. It yields whetstones or scythe-stones of good quality. In the south of Wiltshire the Upper Greensand is about 150 feet thick, at Didcot 70 feet, and in Buckinghamshire it dies out altogether, passing laterally into the clay of the Upper Gault.

In Bedfordshire and Cambridgeshire the Gault outcrops over a considerable area, always forming low ground, and underlying the eastern part of the Fens. Consequently it is largely concealed by a thick cover of river gravels and alluvium. Where exposed at the surface it forms a heavy cold wet soil, of much the same quality as the Oxford Clay. It always needs draining, but owing to the low levels this operation is often difficult to carry out. From recent observations it appears that certain areas near Cambridge formerly mapped as Gault are really covered by a grey clayey or marly deposit. Although strongly resembling Gault in appearance this clay contains fragments of flint derived from the Chalk. It is doubtless largely made up of material derived from the Gault and other formations, laid down in lakes of Pleistocene, possibly late Glacial age. The true Gault soils of this area are best suited for permanent pasture, although there is a good deal of arable land. Where gravels cover the Gault a good deal of fruit is grown.

The Chalk. From the agricultural point of view the Chalk is undoubtedly one of the most important of British rock-formations. It forms a large proportion of the surface area of England in the south and east, and owing to a variety of causes Chalk soils are often more highly farmed than their actual agricultural value would seem to warrant. This state of affairs is brought about by favourable climate, nearness to large markets and other factors.

The Chalk itself is a remarkably uniform formation, thus contrasting strongly with the strata below it. Local variations in the soil are due to superficial deposits of various kinds masking the character of the Chalk. Almost everywhere the Chalk follows directly on the beds of the Gault or Upper Greensand; only in Cambridgeshire is there any real break in the succession (see p. 102). Where the Upper Greensand is not developed the top of the Gault becomes more and more calcareous till it passes into the Chalk Marl, the lowest zone of the Chalk. In the south of England the base of this division is marked by the so-called Chloritic Marl, a layer of soft calcareous sandy clay with glauconite and phosphate nodules. Where the Upper Greensand is found the transition is naturally more abrupt.

The Chalk Marl is soft and forms low ground, but the rest of the Chalk nearly always stands up as conspicuous hills, such as the North and South Downs, Salisbury Plain, the Chiltern Hills and the Wolds of Lincolnshire and Yorkshire. The Chalk hills are lowest between Hertfordshire and the Wash;

this is perhaps due to reduction in height by ice during the glacial period.

The Chalk is generally subdivided into three series, Lower, Middle and Upper. Still smaller subdivisions are partly lithological and partly founded on fossil zones. However the lithological variations are but slight, being for the most part only slight differences of colour and hardness. The oldest classification is founded on the relative abundance or absence of flints, as follows:

Upper Chalk with many flints,
Middle Chalk with few flints,
Lower Chalk without flints.

This classification is not universally applicable, since there are many local variations in this respect, though it is useful as a rough generalization. The flints derived from the Chalk are of much importance in all the succeeding stages of the British formations, at any rate in the south and east of England, since they form the source of the material of a very large proportion of the Tertiary strata, and of the recent superficial deposits. In all of these flint gravels are very common. For an account of the origin of flints see p. 99.

In the south-east of England the Lower Chalk is about 200 feet thick, but it decreases towards the west, and in Devonshire it is represented by 3 or 4 feet only of gritty pebbly sandstone, evidently laid down in very shallow water. In the south of England the Middle Chalk also averages about 200 feet; both divisions also become thinner towards the north. Although the highest zones of the continental Chalk are not found in this country, nevertheless the Upper Chalk is at least 1300 feet thick in Hampshire, and about 1000 feet at Norwich.

The Chalk may be described in general terms as a remarkably pure limestone, sometimes containing as much as 98 per cent. of carbonate of lime. The lower part however is often much less pure than this, the Chalk Marl in particular containing enough clay to make good cement. The Chalk varies in colour from pure white to various shades of pale grey and pale yellow, or even pinkish. The rock is generally much divided by irregular joints, and varies a good deal in hardness; the Chalk

of Lincolnshire and Yorkshire is much harder than that south
of the Wash, while the Chalk of Antrim is the hardest of all[1].

Perhaps the most important practical variation in the Chalk
is in relation to its phosphoric acid content. In some places
only the merest trace of this substance is present, but in a few
localities it is sufficiently abundant to produce a marked effect
on the fertility of the soil, and even to yield substances of manurial value. The peculiar phosphatic deposit of the Cambridge
Greensand has already been described (see p. 102). This
however is quite exceptional, being the direct result of local
uplift and unconformity. Under ordinary circumstances the
phosphate occurs either disseminated throughout the Chalk or
as ill-defined lumps and nodules not differing much in appearance from the rest of the rock. In the well-known phosphatic
Chalk at Taplow the phosphate is due to the presence of fish-
remains in great abundance. A greyish brown phosphatic Chalk
is also found at Ciply in Belgium. (For further details see p. 103.)

The distribution of the Chalk can be most conveniently
described by taking Salisbury Plain as a starting point. From
this centre Chalk hills radiate in four directions: (1) towards
the south-west into Dorset and the east of Devonshire;
(2) slightly south of east into Hampshire and Sussex, ending in
the South Downs at Beachy Head; (3) nearly due east, diverging
from the last and running through the North Downs to Dover;
(4) towards the north-east, as far as Suffolk—in Norfolk the
strike becomes due north, in Lincolnshire and south Yorkshire
north-west. Finally the outcrop of the Chalk swings round to
the east, ending just north of Flamborough Head. There are
also small isolated patches in the Isle of Wight and in the Isle
of Thanet. Each of these six areas possesses its own agricultural
characteristics, and must be separately described.

Perhaps the most typical of all Chalk areas is the South
Downs; here the Chalk is quite free from transported material,
unlike many tracts further north, and true Chalk soils are
found, giving rise to a characteristic landscape, specially

---

[1] The Chalk of Antrim represents only the Upper Chalk of England. The
Irish equivalents of the Lower and Middle Chalk are soft glauconitic sandstones, the Hibernian Greensand.

remarkable from the absence of running streams. A few rivers rise in the Weald and cut across the Downs, but with this exception all the valleys are dry (see also p. 124). The open Downs are generally covered by a thin layer of red flinty soil, a true residual soil, representing the insoluble residue of the Chalk. This soil is deficient in lime, so much so as to require calcareous manures. In places where there has been less solution the soil is pale grey or white and very calcareous, and admirably adapted for sheep-farming, this being the staple industry. The higher parts are generally open grassy pasture, with arable land in the hollows, where the sheep are folded for fattening.

On the North Downs a very large proportion of the Chalk is covered by the clay-with-flints, which seriously alters the character of the soil (see p. 114). Where free from this deposit the Chalk soils are of much the same kind as in the South Downs, but in Surrey and western Kent the character of the farming is largely controlled by proximity to London, from whence also large supplies of manure are obtained. Hence arable land is more abundant than pasture, and the strip of country along the north slope of the North Downs is very highly farmed, especially for potato-growing and dairying. The Chalk area of the Isle of Thanet is almost entirely under the plough, without trees or hedges.

The Down type of scenery is also to be seen over a wide extent of country in Hampshire, Wiltshire, Dorsetshire and Berkshire. Originally almost or quite free from forest, this was one of the first districts to be cultivated regularly on a large scale, and has long been devoted to sheep-farming. Taking Salisbury Plain as an example, the soils on the Chalk Marl in the lower parts of the valleys are generally heavy, and often used as water-meadows. The slopes of the hills are commonly arable land, with open sheep-walk at the top. The arable soils are free-working loams, rather heavy below, light and thin in the higher parts, and often very flinty. The highest arable soils are thin (often only 4 or 5 inches), black and crowded with small flints, with pure Chalk immediately below, there being no real subsoil. The open land is all grass, with the characteristic short sweet herbage of the Chalk. The farms are usually laid out in long strips running down the hill sides, so as to include a fair proportion of all kinds of land.

In Dorset the Chalk soil is usually a yellowish flinty loam, the lower slopes being arable and the uplands grass. On the high ground about Winchester the soil is of the common downland type, grey or brown in colour, light and loamy, often forming a mere skin of turf over the Chalk; it is chiefly in pasture, with some woods. On the lower slopes the soil is generally a red-brown loam, often rather heavy, but thin. Near Basingstoke the soils vary a good deal, being chiefly loams or marls, often stony and thin; the Chalk here carries many large patches of clay-with-flints which yield a much heavier reddish brown soil. The Chalk Downs of Berkshire are mostly under natural herbage, while the beech woods of the Chiltern Hills are a well-known and characteristic feature. In Hertfordshire and Cambridgeshire the Chalk is largely covered by glacial gravels and boulder-clay. Where free from these deposits the Chalk yields a thin white or grey loamy soil, nearly all under the plough, and carrying large areas of turnips and of sainfoin. Grass land is scarce, although there are a few areas such as Newmarket Heath, covered with the natural short turf of the Down type. In Suffolk and Norfolk the Chalk is generally buried under a thick cover of glacial and other superficial deposits; true Chalk soils are rarely to be found in these counties. On the Lincoln Wolds the soils are usually very thin, with little or no subsoil; there are no open downs and little permanent pasture of any kind. The Yorkshire Wolds again form a typical sheep, turnip and barley district. The soil is a light loam, flinty in places, but very free from superficial deposits; the land is nearly all arable, and both fields and farms run large. In the Wold valley which cuts through the centre of the district from Malton towards Driffield the soils are deeper and richer and well suited to wheat. Much of the higher parts of the Wolds were brought under the plough in comparatively recent times, having formerly been largely rabbit warrens and open sheep walks. Chalk also underlies the whole of the plain of Holderness, but is too deeply buried under drift to have any influence on the character of the soils.

CHAPTER XV

THE TERTIARY SYSTEMS

The Tertiary rocks are by most authorities divided into four separate systems, Eocene, Oligocene, Miocene and Pliocene. This classification leaves somewhat doubtful the position of the deposits later than the Pliocene, which are commonly divided into Pleistocene and Recent; by some writers these are regarded as included in the Tertiary, while others introduce another term, Quaternary. Without entering into a discussion of the merits of these classifications it will be sufficient to state here that the strata from Eocene to Pliocene inclusive are included in this chapter, while the Pleistocene and Recent deposits are treated in a separate chapter. This course is sufficiently justified by the great importance from an agricultural standpoint of the latter groups.

Between the deposition of the highest beds of the Chalk seen in Britain and the lowest beds of the Eocene a long period of time must have elapsed and there was also a striking change in the physical conditions. The fairly deep and clear sea of the Chalk gave way to shallow water and estuarine conditions; the greater part of the area was elevated into land and near its shores beds of gravel, sand and mud were laid down; in the north-east of Ireland and in the west of Scotland there was great volcanic activity in the earlier part of the period.

Throughout the greater part of Tertiary time there were important earth-movements in nearly all quarters of the world, leading to the formation of great mountain-chains; most of the important mountain-ranges of the world date from this time. The effect of these movements was felt in nearly all parts of the British Isles, though not so strongly as on the continent; the Jurassic and Cretaceous rocks were tilted and in parts of the

south of England strongly folded and most of the existing
structure and relief of the British Isles date from Miocene
times, when, as it appears, our present river-systems were
chiefly initiated.

Between the Cretaceous and the Tertiary there is also, in
Britain at any rate, a great palaeontological break. Very few,
if any, of the Cretaceous species of fossils are found in the
Tertiary, and the fauna of the latter has a distinctly modern
appearance. The existing species of shells, for example, begin
to appear in the Eocene, and in the Pliocene nearly all the
invertebrates belong to still living species. In the Tertiary
strata also we find the ancestors of the present vertebrates,
including those of our domestic animals; this subject is dealt
with in Chapter XVII.

It should be stated however that the stratigraphical break
between Cretaceous and Tertiary is not everywhere so apparent
as in Britain. In the Mediterranean region and in the United
States, out of many other examples, there are strata clearly
intermediate between the two; in Britain both the Cretaceous
and the Eocene are somewhat exceptional in character, belonging
to inland sea and shallow water facies, whereas, as a rule it is
only in the open sea that perfect continuity is found, as occurs
in the deposits of this age in the south of Europe and northern
Africa.

I. The Palaeogene System. Eocene and Oligocene
Series

Of the subdivisions of the Tertiary the Eocene alone covers
an extensive area in the British Isles. There are now two
isolated patches of considerable size, known as the London
basin and the Hampshire basin respectively. The London
basin forms a triangular area extending from the eastern border
of Wiltshire to the North Sea; on the north it extends as far
as Ipswich while its southern limit is at Deal, in Kent. The
subdivisions generally recognized are as follows:

Bagshot Sands,
London Clay,
Lower London Tertiaries.

The lowest division is thickest in the south and thins to the north. It is usually further subdivided thus:

> Blackheath and Oldhaven beds,
> Woolwich and Reading beds,
> Thanet Sands.

At the base of the Thanet Sands there is generally a peculiar bed of green-coated flints; it is supposed that after the sands were deposited solution of the Chalk went on and these flints were left behind. For this reason also the base of the Thanet Sands is very irregular and they have often fallen into deep pipes formed by solution in the Chalk. The upper part of the series consists of sands and flint-gravels with exclusively marine shells.

The Woolwich and Reading beds consist partly of sands and partly of clays, the latter being dominant in the west and north. The Blackheath and Oldhaven beds, composed of sands and gravels are thin and quite unimportant. The total thickness of the whole of the Lower London Tertiaries is generally less than 100 feet; their chief importance lies in the fact that in many places, especially to the north of the Thames valley, they occur as small outliers resting on the Chalk, and they have undoubtedly supplied material, mainly flints, to many of the superficial deposits mentioned in the next chapter (see also Chapter v). In some localities certain beds have been cemented by silica to form hard and massive grits and conglomerates, including the Sarsen stones of the Wiltshire downs and the Hertfordshire Pudding Stone. The distribution of the Sarsen stones, as for example near Marlborough and on Salisbury plain, shows that the Eocene beds once extended much further than they do now, having been extensively removed by denudation from the elevated ground. In the Hampshire basin only the Reading beds are present, so that deposition must have begun somewhat later in this area, but as shown later the two now separated areas were once continuous.

The London Clay covers by far the greater part of the London basin and is also found in the Hampshire basin. At its thickest it comprises about 500 feet of stiff clay, dark blue in colour when obtained from deep borings and wells, but

weathering brown near the surface. It is of estuarine origin
and contains a mixture of marine shells and land plants; the
latter especially are of an almost tropical character, including
palms and magnolias. Towards the west the clay becomes
more sandy, as also in Sussex and Hampshire. It was clearly
deposited at and near the mouth of a large river that flowed
from an extensive land area in the west.

After the deposition of the London Clay the conditions in
the northern and southern parts of the area of deposition show
a considerable difference; in the Thames valley the upper
Eocene strata consist of a considerable thickness of sands, the
Bagshot Sands, which form the soil of the well-known heathy
barren country about Aldershot, Bagshot, Woking and Ascot.
There are also several small outliers capping the hills at Harrow,
Hampstead and Highgate in the northern suburbs of London,
together with some patches in Essex (Epping Forest). The
maximum thickness is about 150 feet and fossils are scarce. In
Hampshire and the Isle of Wight on the other hand the upper
Eocene strata consist of clays and sands with occasional pebble
beds, containing as a rule abundant marine fossils and some
fresh-water beds with plants. Some parts of this series strongly
resemble the Bagshot Sands, and there is no doubt of their
general equivalence. The fossils, both shells and plants,
indicate a warm sub-tropical climate.

In the Isle of Wight and in the New Forest the Eocene
strata are succeeded conformably by the Oligocene series, a
variable succession of sands, clays and limestones, mainly of
fresh-water origin, and containing abundant land and fresh-
water shells, with some shallow-water marine bands. The total
thickness is 600 or 700 feet, but the area covered is small and
the whole series is of little agricultural importance, the New
Forest being largely unenclosed and covered with heather and
pine woods. Oligocene strata are not found in any other part
of the British Isles.

As before stated there was during Eocene and probably also
Oligocene times great volcanic activity in the north-western
part of Britain, in the north-east of Ireland and the west of
Scotland; this was an accompaniment of the great crust

disturbances of that period, which in Britain culminated in
the Miocene. The volcanic eruptions belonged to a special type
known as fissure-eruptions; there were no volcanic cones, but
the basaltic lava welled out from great cracks in the crust, and
accumulated to a depth of two or three thousand feet in
successive flows, with occasional sedimentary deposits and old
surface soils between the flows, testifying to a considerable
lapse of time. At a later stage great laccoliths of granite and
gabbro were intruded into the lavas, as well as a series of dolerite
sills and dykes. Though originally of much wider extent than
now, the basalt plateaux cover considerable areas in Antrim,
Mull and Skye, while intrusions of this age are also found in the
Mourne Mountains and in Arran, Rum, Eigg and other islands
of the Inner Hebrides. Some very large dykes in southern
Scotland and northern England also belong to this period.

II. The Miocene

The very limited distribution of Oligocene strata in Britain
shows that the greater part of the area was already land at
this time, and during the succeeding period this state of things
continued in an even more marked degree; consequently Miocene
strata are entirely absent. This period was almost everywhere
in western Europe a time of crust-disturbance, denudation and
continental conditions. The tilting and folding of the strata
are very conspicuous in the south of England. From this time
dates the anticlinal uplift of the Weald, which in its westward
continuation into Hampshire and Wiltshire brought up to the
surface a broad band of Chalk and separated the Eocene strata
into two distinct synclinal basins, one in the Thames valley
and the other in Hampshire and the northern half of the Isle of
Wight. Another fold parallel to the Wealden axis affected the
Cretaceous rocks of this island and of southern Dorset. It
forms a sharp monocline and the Chalk and lower Eocene strata
are now almost vertical, as can be well seen at the western end
of the Isle of Wight. The Miocene uplift also gave rise to the
dominant south-easterly and easterly dip of the Mesozoic rocks

of England as far north as Yorkshire; the structure of the
country then became essentially as it now exists.   The Wealden
axis underwent a further slight uplift somewhat later, but apart
from this there has been little differential movement, although
some variation has taken place in the relative levels of land
and sea.

The deposits generally regarded as typical of the Miocene
of western Europe are the Faluns of Touraine.   These are
shelly and marly sands, formerly used to a considerable extent
for spreading on the land as a fertilizer.   They are only of
interest from their strong resemblance to the Pliocene beds of
East Anglia, to be described in the next section.

III.   THE PLIOCENE

At the close of Miocene time the sea again encroached on
the land over a part of southern and eastern England, from
Sussex to Norfolk, and also in Cornwall and South Wales.
South of the Thames deposits of this age are very scanty, and
another uplift of five or six hundred feet carried some patches of
early Pliocene gravels and sands to the top of the North Downs
in Kent, where some small remains of them are still preserved
in pipes in the Chalk.   In Cornwall, South Wales and Wexford
also a few patches of gravel still remain on the surface of the
uplifted plain of marine denudation.

It is only in Essex, Suffolk and Norfolk that Pliocene strata
cover any considerable area and even these are for the most
part masked by glacial sands and gravels, so that their importance as subsoils is slight.   The Pliocene strata of East Anglia
consist in the main of sands and gravels often very rich in shells,
whole or broken, and strongly resembling the Miocene Faluns.
To these deposits the local agricultural term " Crag " is commonly
applied.   The total thickness may be a little over 300 feet, but
the whole series is never developed at any one place; on the
contrary the beds are shown to be successively newer when
followed from south to north, from Walton-on-the-Naze to
Cromer.

The whole succession may be generalized as follows:

Cromer Forest beds,

Weybourn Crag,

Chillesford beds,

Norwich Crag,

Red Crag,

Coralline Crag.

The Coralline Crag, a yellow shelly sand, is only known as a small patch near Aldeburgh. The Red Crag extends over an area of some 300 square miles, but is rarely seen at the surface owing to a covering of glacial deposits. It consists of reddish ferruginous sands and gravels with very abundant shells, often much false-bedded, and clearly formed as sand-banks along the coast line of a sea that was retreating to the north. The Norwich Crag spreads over an area of some hundreds of square miles in eastern Suffolk and Norfolk; it consists of sands, clays and gravels, paler in colour than the Red Crag. A boring at Lowestoft passed through 180 feet of it. The Chillesford beds are of somewhat different character, being river deposits, and it is believed that they were formed in a northward continuation of the Rhine; they cover only a very small area. The Weybourn Crag and the Cromer Forest beds are only seen on the coast. They are also river deposits and the Forest beds contain the stumps of many drifted trees, which once formed "snags" in the river.

When the shells of the Crags are examined in detail certain very important conclusions can be drawn, throwing much light on the climatic and geographical conditions that then prevailed. The great majority of the shells belong to still existing species, the proportion of such increasing upwards. Furthermore in the earlier Pliocene strata there are many shells of species now found in the Mediterranean region. As we rise in the succession the number of these diminishes and they are gradually replaced by more and more northern forms. At the end of the period it is clear that the climate was thoroughly arctic, foreshadowing the approach of the glacial period that immediately succeeded. The connexion with the Mediterranean region was apparently lost, being replaced by free communication with the Arctic

Ocean. This change of geographical conditions undoubtedly
assisted the advent of the northern ice in this country. In
Norfolk especially the Pliocene deposits are succeeded by the
Pleistocene without visible break, and it is not altogether easy
to determine where the line of demarcation should be drawn.

In the few places where the Crags form the actual surface,
free from glacial deposits, they yield a light type of soil, usually
of very poor quality.

CHAPTER XVI

THE PLEISTOCENE AND RECENT FORMATIONS

Owing to the absence of the rest of the Tertiary strata over the greater part of this country, the Pleistocene and Recent formations usually rest with a decided unconformity on older rocks; as before mentioned it is only in Norfolk that there is any real transition. This subdivision includes all deposits formed between the end of the Pliocene and the present day. It is usually divided into two stages, Glacial and post-Glacial, but this is unsatisfactory since the Glacial period came to an end in different places at different times, and in the higher mountains of Europe and within the Arctic circle it is still going on. Hence the terms glacial and post-glacial will here be used only in a genetic sense, not chronologically, it being understood that the later glacial deposits of one area, e.g. Switzerland or Norway, are equivalent in time to the post-glacial deposits of Great Britain or of northern Germany.

The deposits of this age comprise what is known collectively to the Geological Survey as *drift*. An older and still more inappropriate name, *diluvium*, still lingers on the continent, though long abandoned in Britain. It is the indication of the drift deposits in this sense that constitutes the difference between the "solid" and "drift" editions of the maps of the Geological Survey. The only exception is that certain great spreads of alluvium and peat are inserted in the solid maps, owing to the absence of exposures and the impossibility of ascertaining what lies beneath them.

The general characters of the glacial deposits have already been described in the chapter on superficial deposits (Chapter v).

It remains here to give some account of their distribution in the British Isles.

It has already been stated that much difference of opinion still exists as to the exact manner in which the lowland glacial deposits of Britain were formed. It is admitted by all that valley glaciers of the Alpine type existed in the mountains of Scotland, northern England, Wales and Ireland. The main controversy is concerned with the genesis of the boulder-clays, gravels and sands of the lower-lying districts and especially with those of the Midlands and eastern England. The most striking feature is the existence in these of innumerable boulders of far-travelled rocks, that must have been brought from northern England, Scotland and Norway. The last are of special interest, since their origin is so well-established. Owing to the distinctive character of many of the Norwegian rocks there can be no possible doubt as to their source. The unmistakable igneous rocks of the Christiania district, and especially one variety known as "rhomb-porphyry," have been found all over eastern England, in Yorkshire, Lincolnshire, Norfolk, Cambridgeshire, as far south as the northern suburbs of London, and as far west as Bedford. Hence it is clear that Norwegian ice advanced to the coast and penetrated far up the valleys of the Midland rivers. Associated with these boulders are also many from Scotland and northern England, including abundant blocks of Carboniferous Limestone and Millstone Grit, as well as igneous rocks of many kinds. On the western side rocks from south-western Scotland and the Lake District are abundant in Cheshire, Staffordshire and Shropshire, while Welsh boulders extend as far east as Birmingham. Another striking fact is the transport of boulders of granite and other rocks from the Lake District over the Pennine chain into east Yorkshire, both down the Tees to the coast and down the Vale of York. Again, northern rocks are found on the north coast of Wales and Scotch rocks in Ireland; hence it is clear that the Irish Sea also was filled with ice coming from the north, which travelled over the plains of south Lancashire and Cheshire, against the natural drainage of the country. The valleys of central and south Wales possessed their own glacier systems, which followed

in the main the natural drainage lines, with occasional diver-
gences. More difficult to comprehend is the state of affairs in
the Midland counties; even yet it has not been decided with
any certainty whether glaciers existed in the hills of south
Yorkshire and Derbyshire. However certain anomalies in
boulder-distribution are most easily explained on the hypothesis
of a late development of glaciers in this region after the partial
withdrawal of the sea-ice. It appears that at a late stage there
was a transport of boulders across Norfolk and Suffolk from
north-west to south-east, and a similar effect can also be traced
in Rutland and Lincolnshire. This was probably due to a
Pennine glacier. Boulder-clays and other glacial deposits are
abundant in Leicestershire, Northamptonshire, Warwickshire,
and almost as far south as Oxford. These appear to be of
complex nature, due to ice-streams coming from different
directions. Over a large part of this area Chalk and flints are
common in the drifts and the influence of the North Sea ice
seems to have extended far to the west, though its exact limits
are as yet unknown.

In Scotland the relations are somewhat more simple. In
general terms, glaciers moved down all the great valleys towards
the sea, overrunning the lower lands and leaving thick deposits.
The Isle of Skye had a small independent ice-cap, with its centre
in the Cuillin and Red Hills. The behaviour of the eastward
flowing glaciers of the mainland, when they encountered the
North Sea ice, is still a matter of uncertainty. The ice from the
Moray Firth seems to have been driven north-westwards over
Caithness, but south of Aberdeen the evidence is conflicting.
Rocks from the valley of the Forth, near Edinburgh, have been
identified in considerable numbers in Norfolk and Cambridgeshire and probably some from Forfar and Aberdeen. However
some Scotch geologists maintain that the ice from the Forth
and Tay chiefly went northwards.

There was a great development of glaciation in Ireland, the
structure of this island being very favourable to the accumulation of an ice-sheet, since most of the mountains are near the
coast, with a low plain in the middle. The conditions must
have been very similar to those now prevailing in Greenland.

The general movement was radially outwards in all directions except in the north-east, where the local ice was overpowered and driven back by that coming from Scotland.

The glacial deposits consist of clays, sands and gravels; these vary in their constitution and character according to the source whence they are derived. The boulder-clays show wide variations both in the nature of the included blocks and in the finer matrix. In most cases it is clear that the greater part of the material is of local origin, having been derived from the region passed over by the ice just before reaching the spot where the examination is made. Thus in eastern England the constituents of the boulder-clay were evidently derived chiefly from the north-east or north; in the western Midlands chiefly from the north-west and so on. In the glacial deposits of the mountainous regions the clay as well as the stones have come from the higher parts of the country. Since argillaceous material is often somewhat scarce in mountains, the boulder-clays of such regions are often very stony indeed, partaking more of the nature of coarse angular gravels; in fact they are just like the moraines of existing glaciers. A somewhat more difficult problem is the origin of the red and purple stony clays of Lincolnshire and Yorkshire. It is probable that they are largely composed of Triassic material derived from the floor of the North Sea. The Triassic strata reach the coast in north Yorkshire and we have no means of ascertaining how far they extend to the eastward of the present coast-line. Fragments of Lias and other Jurassic rocks are very numerous in some of the eastern drifts, and the very abundant Chalk was derived from the Lincolnshire Wolds and from the neighbourhood of the Wash. There can be no doubt that the Chalk hills of south Lincolnshire and of Norfolk were much reduced in height by the passage of the ice and before the glacial period the Wolds of Lincolnshire probably extended further south than now. The pebbles of Chalk so common in the clays and gravels of Norfolk and Cambridgeshire are much harder than the local Chalk, resembling that of Lincolnshire and Yorkshire in this respect. In East Anglia also are many tabular grey flints of Lincolnshire type, quite unlike the local black flints. In the drifts of Cheshire and Lancashire are many well-rounded sand-grains derived from the Trias sandstones.

On the whole organic remains are not very abundant in the drifts; marine shells, whole or broken, are found in some gravels in Yorkshire, Lincolnshire and elsewhere, near the eastern coast. Of more interest is the occurrence of marine shells in some places in North Wales and Cheshire at considerable heights. The most famous of these occurrences is at Moel Tryfan in Caernarvonshire, where some sixty species of shells have been found on a mountain top, nearly 1400 feet above sea-level. These have been brought forward as proofs both of the land-ice and of the submergence theories. Some teeth of elephants and bones of other animals found in eastern England have been derived from Pliocene deposits.

With the evidence at present at our disposal it is impossible to decide the vexed question whether the whole glacial history of the British Isles was comprised in one single advance and retreat of the ice-sheet, as was at least tacitly assumed by many writers till within very recent years, or whether there have been several glaciations separated by interglacial periods of mild climate. In Switzerland it has been shown that there were four separate advances with intervening periods when the climate was milder than at present. In Norway also two or even three such periods are demonstrated. It is to say the least very improbable that Britain wholly escaped these fluctuations, but the occurrence of warm interglacial periods is not yet conclusively proved. At any rate it is well known that in some parts of England there are two or even three boulder-clays in vertical succession, derived from different sources, besides a variety of sands and gravels, as in the east of England, from Yorkshire to Suffolk. The clearest evidence of climatic fluctuations is seen in the peat-mosses of Scotland (see p. 119), but it has not yet been found possible to correlate these with the glacial deposits of England. We do not yet know to what extent man was contemporaneous with the glacial period in Britain; a completely glaciated country is obviously unfit for colonization, but during intervening periods of mild climate, if such occurred, the country was probably inhabited. There seems no doubt at any rate that man lived in Europe during the earlier glaciations of the Alps, and that Great Britain was then joined to the continent, so that migration might have occurred.

The southern limit of true glacial deposits in England coincides approximately with a line drawn from the mouth of the Thames to the Bristol Channel, with a bend to the north in the neighbourhood of Oxford. South of the Thames large areas of the country are quite free from drift, though large spreads of gravel and alluvium are often found. Some of these must be equivalent in age to the glacial drifts of the north. One of the most important types is the Coombe Rock or Head, already described (see p. 124). The gravels of this region are also described in the chapter on superficial deposits (p. 125). North of the Thames almost the whole country is masked by a more or less thick covering of glacial drift. Only a very few areas of any considerable size are completely free from drift; ·these comprise the hilly parts of east Yorkshire, and the lower part of the Vale of York, continuing southwards up the Trent valley about as far as Nottingham. It is manifestly impossible to give a full account of all the types of glacial deposit found even in the lowland and cultivated parts of the British Isles, and a few selected examples must suffice.

Yorkshire and Lincolnshire. The drifts of eastern Yorkshire form a somewhat complicated series, and the true relationship of the different members is not easy to ascertain. In the northern part of the county the most conspicuous is a reddish brown boulder-clay of quite remarkable tenacity. This spreads widely over the low ground, and fills up the seaward ends of many valleys. Here and there are found in it seams of sand or gravel and boulders of all sizes are abundant, especially those from Scotland and the Lake District; particularly notable are the great boulders of Shap granite from Westmorland, sometimes over a ton in weight. This clay gives rise to a very heavy soil, much of it being rather boggy pasture; when well drained it is fertile and can carry heavy crops. South of Flamborough Head, in the low-lying district of Holderness, several distinct clays can be recognized, overlain in some places by extensive spreads of gravel. The total thickness of all the deposits is about 100 feet. The boulder-clay forms a stiff rich marly soil, which however, owing to special local causes, varies somewhat in texture from place to place. The more elevated parts of the boulder-clay are covered by a loamy soil, not specially heavy, and Mr Clement Reid[1] attributes this fact largely to the action of the wind which has blown away the finer particles of clay, thus lightening the soil. This fine material is again deposited in more sheltered localities, thus making the clays locally still heavier and stiffer, but occasionally improving sandy patches (see also p. 139). The boulder-clay soils of Holderness are naturally good fertile wheat lands, but of late years much has been laid down to grass. The alluvial deposits of this area are described elsewhere.

Throughout the great central plain of Yorkshire, from the Tees to a little south of York, there is a vast spread of glacial deposits. Along the Tees two clays can be recognized, separated by sands and gravel. The lower boulder-clay is bluish when fresh and contains many foreign boulders, often well striated; it rises high on the flanks of the Cleveland and Hambleton Hills. The upper boulder-clay is reddish, with few stones, and it does not ascend nearly so high.

In Lincolnshire there are two quite distinct areas of boulder-clay. On the east of the Wolds are found two clays, as in Yorkshire; the purple clay below and the Hessle clay above, with occasional beds of sand and gravel. The purple clay is a stiff purplish-brown clay with many foreign boulders, while the Hessle clay is greyish-brown and sometimes contains a little Chalk. On the west of the Wolds most of the low country is covered by the chalky boulder-clay, whose character is perhaps sufficiently explained by its name; sometimes at first sight this looks like Chalk in place, but enclosed fragments of other rocks soon show its true nature. It forms part of the great sheet of chalky boulder-clay that covers so much of the eastern and midland counties almost as far south as the Thames.

East Anglia. The general succession of the glacial deposits

---

[1] Reid, "Geology of Holderness," *Mem. Geol. Survey*, 1885, p. 115. "Dust and Soils," *Geol. Mag.* 1884, p. 165.

of Norfolk, Suffolk and Cambridgeshire has already been described (p. 131). Of the divisions there enumerated only two, the glacial sands and gravels, and the chalky boulder-clay, are widespread and of agricultural importance, the others being local. The soils derived from the glacial formations of this district show an enormous amount of variation, ranging from the barren and uncultivated sands of the Breckland of western Norfolk to heavy clays in Cambridgeshire. The problem is also much complicated in many parts by the occurrence of different kinds of brick-earths and silts laid down in still water in late glacial times; the exact manner of origin of many of these is still a mystery, since they do not always occur on the lowest ground, being found also on ridges elevated above the general level of the surrounding country. Some of these were formerly mistaken for boulder-clay, Gault clay and other formations.

In Suffolk the boulder-clay occupies a very large area; it is generally very chalky, but varies much in character, ranging from good corn land to poor heavy clay. Most of it is well suited to wheat and beans, the lighter varieties to barley. The glacial sands and gravels vary from land capable of producing excellent crops of barley to light hungry heath land, almost worthless from the agricultural point of view.

The Western Midlands. Another area of glacial deposits of great agricultural importance forms the wide plains of south Lancashire and Cheshire, with extensions into North Wales, Shropshire and Staffordshire. Most of this country is underlain by Trias, and this formation has an important influence in determining the character of the drifts on the principle already explained. The ice apparently moved inland from the Irish Sea, ploughing up the soft Triassic sandstones and marls and incorporating with them many boulders from the Lake District and southern Scotland. The result is a boulder-clay of a pre-vailing red tint, with a strong resemblance to the local Triassic deposits. This forms heavy soils, mostly under grass and devoted to dairying. The glacial sands and gravels on the other hand generally occupy somewhat higher ground, and form light sandy soils, often with a *pan* (locally called Fox-bench) at a depth of 4 to 14 inches. This is clearly analogous to the Ortstein of German writers. The high-level shell-bearing glacial gravels of North Wales and Cheshire have already been mentioned.

The boulder-clays and glacial gravels of the western Midlands are of very varying character and origin. Boulders from North Wales are found as far east as Birmingham, while rocks from the Lake District are abundant in the middle part of the Severn valley, about Shrewsbury. All over the western Midlands the drifts contain in abundance the characteristic rounded stones from the Bunter Pebble Beds of Staffordshire and Cheshire, and these extend about as far south as Oxford. The chalky boulder-clay of eastern type also appears to stretch westwards into Warwickshire and the line of demarcation between the drifts of western and eastern origin is uncertain. It is impossible, owing to their variety, to give any general account of the drift soils of these regions; they vary from stiff clays to light sands, according to the character of the under-lying or immediately contiguous rocks, which belong to many different formations within a comparatively small area.

In the north of Shropshire drift deposits cover a large area. They include sand and not very stiff clays. Most of the latter yield medium loams of a brownish colour that can grow good wheat and oats, but there is much grass, especially near Shrewsbury, devoted to cattle-farming and dairying. The drift sands form light soils, sometimes peaty and generally deficient in lime; they are specially good for potatoes and carrots.

Northern England and Scotland. Glacial deposits of various kinds are of almost universal occurrence in the valleys and plains of northern England and southern Scotland. During the earlier part of the Pleistocene period the mountains gave rise to local glaciers which left their deposits on the lower ground almost everywhere and greatly modified the agricultural character of these areas. The plain of north Cumberland and the valley of the Eden were overflowed by ice coming from the Galloway district, bringing innumerable boulders of granite and other rocks. These drifts are of great thickness and render difficult the demarcation of the underlying rock-systems. In the Eden valley in particular the peculiar and characteristic forms of the glacial deposits are well marked, especially the long narrow ridges commonly known as drumlins, which lie with their long axes parallel to the direction of movement of the ice. They consist mostly of a very stony boulder-clay, with occasional seams of gravel and sand. Since the solid rocks below belong to the Permian and Triassic systems the glacial deposits are generally of a red colour. Around Carlisle and stretching well into Scotland is a great spread of alluvium, often of a peaty nature, and including the well-known Solway Moss. The lower ground of the Southern Uplands is also for the most part occupied by glacial deposits, formed by the local ice-streams, which radiated in all directions from the higher mountains and especially from the granitic peaks of Galloway. The mixing of soil-material thus brought about has increased the fertility of this region, which is now a district of good farming and stock-breeding; the Galloway and Ayrshire cattle are famous.

In northern Scotland the evidences of glaciation are almost everywhere very conspicuous; the mountains show the characteristic features of ice erosion, and the valleys and plains are occupied by great thicknesses of boulder-clay and other glacial deposits. The boulder-clays, which are often very stony, are interstratified locally with beds of gravel, sand and peat, completely masking the underlying solid formations and yielding soils of the most varied character, generally however somewhat on the heavy side. The richest of these soils are those composed of debris from the volcanic rocks of Old Red Sandstone and Carboniferous age that cover such large areas in the central valley of Scotland. Perhaps best of all are the soils of the Carse of Gowrie, which contain much material derived from the lavas of the Ochil and Sidlaw Hills. In the highly farmed and highly rented district of East Lothian, the soils are derived very largely from boulder-clays resting on strata of the Old Red Sandstone system, most of the material of the glacial deposits being of local origin.

The soils of the important agricultural district of Aberdeen-shire are almost entirely on drift, very few being derived from rock in place. The glacial deposits have however in places been considerably worked over and modified by running water, especially in the river valleys. The most striking characteristic of the soils of this area is their variability. In parts there are heavy glacial clays, while in other places the soils are very light and gritty. Over considerable areas also the soils are peaty. It is very difficult to find any considerable area of uniform type and no useful generalizations can be drawn up.

CHAPTER XVII

THE GEOLOGICAL HISTORY OF THE DOMESTIC ANIMALS[1]

A work on Agricultural Geology could scarcely be considered complete without some attempt to state what is known concerning the origin and geological history of those animals which, by reason of their utility in the service of Man, have ever been a source of interest, especially to agriculturists. The subject is one of great complexity, and is discussed here mainly from the standpoint of well-established facts. Nevertheless the hypotheses that have been put forward are full of interest and have served to draw attention to the importance of the question.

The animals dealt with are the Horse, the Ox, the Sheep, the Pig and the Dog. In the case of some of these species the evidence available for building conclusions is of the scantiest, and they are accordingly dealt with here with corresponding brevity.

THE HORSE

It is generally agreed, even by those who most definitely adhere to the view that our modern horses and ponies have had a multiple origin, that the little primitive four-toed *Hyracotherium*, first described in 1839 by Owen, and found in the London Clay of Studd Bay, Herne, Kent, stands very near to the base of the stem from which all the equine animals subsequently arose. The American palaeontologist, Cope, regarded the still older five-toed and semi-plantigrade *Phenacodus* of the lower Eocene of South America as a more remote ancestor, but this animal was probably too large and in certain respects too

[1] By F. H. A. Marshall, Sc.D.

specialized to have been in the direct line of equine descent. The *Hyracotheriidae* died out in Europe at the close of the Eocene period; at least no representatives of the equine family have been found fossil in the Oligocene of the Old World, and we have to pass to the American continent to trace the progress of the Horse's descent. The record shown in the Tertiary strata of North America is very complete, and consequently the inference has been drawn that the evolutionary process by which the five-toed and four-toed ancestors of the Horse became transformed into the modern single-toed *Equus* took place exclusively, or almost exclusively, upon that continent, the various genera which have been found fossil in the later Tertiary formations of Europe and Asia being supposed to have migrated from the New World. It must however be pointed out that such an inference is of very doubtful validity, since the absence of fossil *Equidae* in the Oligocene strata of Europe and Asia is only negative evidence.

In North America *Hyracotherium* was represented by the closely similar *Eohippus*. The descent of the modern Horse from this genus has been described by the American geologist Marsh, whose researches on equine lineage have become classical. He thus summarizes the results of his investigations:

"The oldest representative of the Horse at present known is the diminutive *Eohippus* from the Lower Eocene. Several species have been found, all about the size of a fox. Like most of the early mammals, these ungulates had forty-four teeth, the molars with short crowns and quite distinct in form from the premolars. The ulna and fibula were entire and distinct, and there were four well-developed toes and a rudiment of another on the forefeet, and three toes behind. In the structure of the feet and teeth, the *Eohippus* unmistakably indicates that the direct ancestral line to the modern Horse has already separated from the other perissodactyles, or odd-toed ungulates.

"In the next higher division of the Eocene another genus, *Orohippus*, makes its appearance, replacing *Eohippus*, and showing a greater, though still distant, resemblance to the equine type. The rudimentary first digit of the forefoot has disappeared, and the last premolar has gone over to the molar

304 THE GEOLOGICAL HISTORY [CH.

series. *Orohippus* was but little larger than *Eohippus*, and in most other respects very similar. Several species have been found [but none occur later than the Upper Eocene.]

"Near the base of the Miocene, we find a third closely allied genus, *Mesohippus*, which is about as large as a sheep, and one stage nearer the Horse. There are only three toes and a rudimentary splint on the forefeet, and three toes behind. Two of the premolar teeth are quite like the molars. The ulna is no longer distinct or the fibula entire, and other characters show clearly that the transition is advancing.

"In the Upper Miocene *Mesohippus* is not found, but in its place a fourth form, *Miohippus*, continues the line. This genus is near the *Anchitherium* of Europe, but presents several important differences. The three toes in each foot are more nearly of a size, and a rudiment of the fifth metacarpal bone is retained. All the known species of this genus are larger than those of *Mesohippus*, and none of them pass above the Miocene formation.

"The genus *Protohippus* of the Lower Pliocene is yet more equine, and some of its species equalled the Ass in size. There are still three toes on each foot, but only the middle one, corresponding to the single toe of the Horse, comes to the ground.

"In the Pliocene we have the last stage of the series before reaching the Horse, in the genus *Pliohippus*, which has lost the small hooflets, and in other respects is very equine. Only in the Upper Pliocene does the true *Equus* appear and complete the genealogy of the Horse, which in the post-Tertiary roamed over the whole of North and South America, and soon after became extinct. This occurred long before the discovery of the continent by Europeans, and no satisfactory reason for the extinction has yet been given. Besides the characters I have mentioned, there are many others in the skeleton, skull, teeth, and brain of the forty or more intermediate species, which show that the transition from the Eocene *Eohippus* to the modern *Equus* has taken place in the order indicated[1]."

In addition to the North American genera of *Equidae*

[1] Marsh, "Introduction and Succession of Vertebrate Life in America," *Nature*, vol. XVI, 1877.

referred to in the above cited quotation, certain horse-like animals have been found fossil in Europe, and these, as already indicated, have been regarded by some authorities as lateral offshoots from the main equine stem, which after attaining a considerable degree of specialization subsequently became extinct. Prominent among these is the *Hipparion* of the Lower Pliocene strata of Germany, Spain, Greece, Persia, China, and other parts of the Old World. In *Hipparion* the lateral digits are rather longer than in *Protohippus*, but they barely extend beyond the level of the first phalanx of the middle digit. The most striking peculiarity, however, is that the anterior inner pillar of the molar teeth is entirely surrounded by enamel, and is thus separated from the rest of the crown, in this respect being markedly different from the condition found in *Equus* and other members of the horse family.

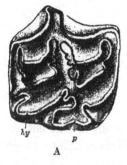

Fig. 44. A left upper molar of a Single-toed (*Equus*), A, and a Three-toed Horse (*Hipparion*), B. *p*, anterior pillar; *hy*, posterior pillar. (From Lydekker.)

Another European genus, but belonging to an earlier epoch, is *Anchitherium* which seems to have been akin to the American *Miohippus*. It has been found fossil in the Lower Miocene fresh-water deposits of Sansan, in France. *Anchitherium* was about the size of a sheep, and the lateral toes though smaller than the middle ones still probably reached to the ground. The yet older *Hyracotherium* has already been referred to.

It used to be supposed by palaeontologists that the horses and ponies which we know to-day were descended by a direct and single line of ancestry from *Eohippus* or *Hyracotherium*.

We now know that at various times during the Tertiary period there were several kinds of horses existing contemporaneously and that in some cases two or more species inhabited the same area. In Miocene times we have the somewhat coarse-limbed *Hypohippus* which was apparently adapted for living in forest areas, and the comparatively slender-limbed *Neohipparion* which was specialized for a desert life, both of these horses having been found in North America. *Hypohippus* is of special interest in that it had a vestige of the first metacarpal bone, such as is still very occasionally found in coarse-limbed breeds of horses but never in slender-limbed breeds. Another Miocene horse, *Merychippus*, was fine-boned, and, according to Ewart[1], bears some affinities to the modern "Celtic pony" or "plateau" horse.

The first to suggest that modern horses had a multiple origin was Hamilton Smith, writing in the middle of the last century, but his views were generally disregarded. In recent years, however, Osborn[2], on the palaeontological side, Ewart, as a result of a careful study of existing breeds and numerous cross-breeding experiments, and Ridgeway, from the standpoint of the archaeologist, have arrived at conclusions which without being identical agree in postulating several independent origins for the domestic horses of to-day.

Ewart, who has paid more attention to this subject than any other zoologist, refers to three different Miocene species of the genus *Equus* which were probably concerned in the formation of existing varieties of horses. These are *E. sivalensis* of the Indian Pliocene, *E. stenonis* of the Pliocene deposits of Europe and North Africa, and a species named by Ewart *E. agilis* (= *Asinus fossilis* of Owen), which has been found fossil in the Pliocene formations of Italy and France, as well as in Pleistocene deposits of North Africa, France, and England (e.g. in a cavernous fissure at Oreston, near Plymouth). These three species agree in having a short anterior internal pillar (a fold of enamel on the inner surface of the tooth) on the premolars

---

[1] Ewart, Appendix to *The Shetland Pony*, by C. and A. Douglas, Edinburgh and London, 1913.

[2] Osborn, *The Age of Mammals*, New York, 1910.

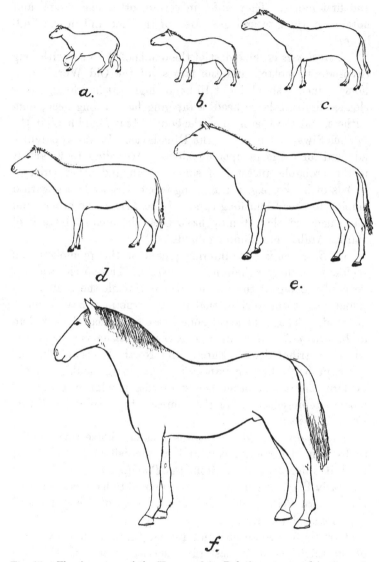

Fig. 45.   The Ancestors of the Horse and its Relatives compared in size and
    form with their typical modern representative.   *a, Hyracotherium* of the
    Lower Eocene; *b, Orohippus,* of the Middle Eocene; *c, Mesohippus,* of the
    Oligocene; *d, Merychippus,* of the Miocene; *e, Pliohippus,* of the Pliocene;
    f, the Horse, *Equus caballus.*   (From Lydekker.)

and first molar. They differ in certain other characters, and
notably in the size and deflection of the face and in the limb
bones[1].

E. *sivalensis* is the oldest of the one-hoofed horses with long
(hypsodont) molars, and the largest of the Old World fossil
horses, since it stood about 15 hands high. It had long, rather
slender limbs, a large head, a tapering face, a long neck, high
withers, and a tail set high on the body. It is found fossil in the
famous Siwalik deposits of the Himalayas. It was apparently
adapted for a life in upland valleys. According to Ewart, it
is the probable ancestor of some Indian and other Oriental
breeds of horses (e.g. certain long-faced Kirghiz horses with a
sloping forehead and long ears). Lydekker on the other hand
has suggested that it may have been the ancestral stock of
Barbs, Arabs, and Thoroughbreds[2].

In *E. stenonis* the anterior pillars of the premolars and
molars were shorter than in *E. sivalensis*. This species was not
generally so big as the Siwalik Horse, though the limb bones
found fossil seem to show that it may sometimes have reached
15 hands. The metacarpal bones were somewhat thicker than
in *E. sivalensis*. According to Ewart, *E. stenonis*, which was
widely distributed over Europe and North Africa in Pliocene
times, probably had an important share in the making of the
modern Shires and other heavy breeds, the latter being the
modern representatives of the "forest" type called by Ewart
Equus robustus.

The *Equus agilis* of Ewart was a smaller horse with slender
limbs. In the metacarpals and in the pillars of the molar
teeth it differed but little from the *Pliohippus* of the Miocene
and early Pliocene periods. It is supposed to have given rise to
the modern Celtic ponies and other horses of the "plateau"
type referred to below.

Coming to Pleistocene and Recent times we find evidence
of the existence of at least three species of horses. Of these

[1] Ewart, "The possible Ancestors of the Horses living under Domestication,"
Proc. Royal Soc. B, vol. LXXXI, 1909.   "The Origin of the Clydesdale,"
Trans. Highland and Agric. Soc. 1911.

[2] Lydekker, *The Horse and its Relatives*, London, 1912.

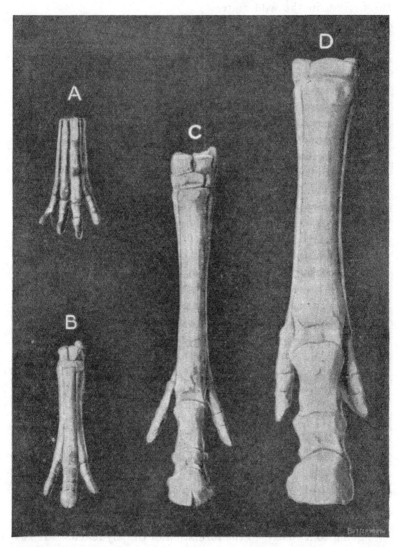

Fig. 46. Bones of forefeet of extinct forerunners of the Horse. A, *Hyracotherium* or *Eohippus*; B, *Mesohippus*; C, *Merychippus* or *Protohippus*; D, *Hipparion*. (From Lydekker.)

one only, the Steppe Horse of Mongolia, or *E. przevalskii*, still survives in the wild state.

This species is characterized by a long narrow face, large nasal chambers, a "Roman-nose," a long pillar on the premolar and first molar teeth, a short back, clean limbs, close hocks, elongated hoofs, large hind-chestnuts, and a mule-like tail set on high up. It is highly specialized for a steppe life. The characteristic colour is a yellow dun. Horses of this type have been found fossil in the Roman camp at Newstead.

In the forest type (*E. robustus*, Ewart) the face is short and broad and in a line with the cranium, the internal pillar on the molars is large, the middle metacarpal is short and wide, the back is long, the hindquarters are rounded, the hoofs are broad, the hind-chestnuts are well developed, and the tail is set on low down. This type is represented among modern breeds by the Shire, the Gudbrandsdal of Norway, and other heavy cart-horse varieties. Forest horses have been found fossil in the Brighton Elephant Bed, in Kent's Cavern at Torquay, and in the much more recent Roman camp at Newstead. The Stone Age deposits at Solutré also contain remains of Forest horses.

In the "plateau" horses the face is fine and tapering, the head is typically small, the pillars of the molars are short, the neck is long and the back short, the middle metacarpals are very slender, the hoofs are small and the hind-chestnuts (hock callosities) and all the four ergots are typically wanting. On the upper part of the tail there is a fringe of short hairs forming a tail-lock, which is shed at the beginning of summer. The typical colour is light dun, and there is often a dorsal stripe or eel-mark and some cross stripes on the legs[1]. The Celtic pony is represented in its present form by some of the Iceland and Faroe ponies and by the Udganger ponies of Norway. It has formed the basis of most of the pony breeds of the world (including such varieties as the Fjordhest of Norway), besides having contributed to the formation of the Clydesdale. It has been suggested that the "Celtic" or "plateau" characters of many modern British breeds are due to infusion of Norwegian

---

[1] Ewart, "The Multiple Origin of Horses and Ponies," *Trans. Highland and Agric. Soc.* 1904.

blood and that ponies of the *agilis* type were brought over not merely to Iceland and the Faroes but also to Britain between the eighth and tenth centuries[1]. According to Ewart *E. agilis* had at one time a wide distribution, for the Arab and the other finely built horses are supposed to have sprung from this stock. Remains of the "plateau" or Celtic variety have been found fossil in Kent's Cavern and other Palaeolithic deposits, and in the Roman camp at Newstead[2].

It remains briefly to consider Ridgeway's views concerning the origin of the breeds from which the Thoroughbred has been derived[3]. According to this author there were two chief varieties which have contributed in varying proportions to the formation of all the improved breeds of horses. The first is the coarse, large-headed horse of Europe and Asia. This animal appears to have been the same as the Forest Horse of Ewart, but bore a relationship to *E. przevalskii*. The second was the African variety called by Ridgeway *E. caballus libicus* which was characterized by a fine head, a star on the forehead and white points, a tendency to striping, a tail set on high up, and the absence of hind-chestnuts. This type is supposed to be represented by the better class of Arabs and Barbs at the present time, and to have been the foundation stock of the Blood-horse and of all the finer types of horses in the world; nevertheless in giving rise to these it underwent varying degrees of admixture with the large, coarse-headed variety referred to above. Concerning the existence of the Celtic pony as a separate sub-species Ridgeway is doubtful, and he appears to regard this animal as probably descended also in part from a Libyan ancestor.

Ewart has pointed out that skulls of the Celtic type have been found at Newstead, and that these closely resemble certain high-caste Arabs, and this fact, together with other evidence, has led him to infer that the Celtic pony, the Libyan horses

---

[1] Marshall, "The Horse in Norway," *Proc. Royal Soc. Edin.* vol. XXVI, 1905.

[2] Ewart, "On Skulls of Horses from the Roman Fort at Newstead," *Trans. Royal Soc. Edin.* vol. XLV, 1907.

[3] Ridgeway, *The Origin and Influence of the Thoroughbred Horse*, Cambridge, 1905

of the second century, and the better type of modern Arab, are all derived from the same ancestral stock.

Without expressing a complete adherence to either of the theories enunciated above, it may be permitted to say in conclusion that there appears to be a considerable body of definite evidence derived from various sources—partly palaeontological, partly historical, and partly experimental—in favour of the view that the various breeds of horses and ponies as we know them to-day have had a multiple origin, and that in all probability more than two species of equine animals have contributed to their formation.

THE OX

There are few data supplied by geological study on which to formulate definite conclusions concerning the origin and development of existing varieties of domestic cattle. Such evidence as is available for this purpose has been procured chiefly from archaeological and historical research, and in view of the meagreness of the information obtained it is not remarkable that the views which have been put forward differ widely on certain points.

According to Zittel the Ox and all other hollow-horned Ruminants, as well as the whole tribe of deer, had a common ancestor in a small Chevrotain-like animal named *Gelotes*, in which the radius and ulna of the foreleg and the tibia and fibula of the hindleg were well-developed separate bones and not unequally developed and united as in the modern Ox or Sheep. *Gelotes*, which is found fossil in Eocene and Oligocene beds, is supposed to be descended from a still more generalized type of Artiodactyl Ungulate, *Pantolestes*, from the Lower Eocene of North America.

In the Pliocene formations several species or varieties of *Bos* have been found fossil, often associated with *Bison*. The earliest known *Bovidae*, like the Chevrotains, were probably hornless, but in the Lower Pliocene remains have been found of cattle which are believed to have been horned in the male, but hornless in the female. *Leptobos falconeri* of India and

XVII]   OF THE DOMESTIC ANIMALS   313

L. etruscus and *L. lutus* of Italy, all Lower Pliocene species, are probably instances of the occurrence of this condition[1]. The remarkably short skull is another characteristic of the genus.

The genus *Bos* appears to have been represented in Pliocene times by two or three species found fossil in the Siwalik Hills of India. Of these *Bos planifrons* (so-called from its flattened forehead) is supposed by Duerst to have been the ancestor of *B. primigenius* and *B. namadicus*, from both of which it differed but little. According to another view *Bos acutifrons* (named by Lydekker after the sharp ridge running longitudinally down the forehead[2]) from the Siwalik Pliocene was the predecessor of *B. namadicus*. It seems probable that all these forms were closely related, and that whereas the sub-species which extended westward became *B. primigenius*, the variety which remained and spread in the East was *B. namadicus*. This species is found fossil in the Pleistocene gravels of the Narbada Valley in Central India. It approximates somewhat to the Gaur and Banting in frequently having the horn-cores compressed at the base.

Bos primigenius, which is identified with the Urus of Caesar, occurs fossil in Pleistocene deposits in the British Isles and various parts of Europe. Its remains have been found in Scottish bogs and East Anglian fens, in alluvial and lacustrine deposits, in association with bones of the Elephant, the Hippopotamus, the Rhinoceros and other mammals. Its first definite appearance was in early Palaeolithic times, when it lived alongside the Bison, though fossil remains which have been referred by some investigators to the Urus have been discovered in the Upper Pliocene of Germany and other parts of Europe. It seems probable that the Urus first arrived in Europe just before the beginning of the Glacial period. The Bison died out at the end of the Palaeolithic period but the Urus survived, for the latter animal is especially characteristic of the Neolithic Age. The following is Nilsson's description of the Urus: "The forehead [is] flat; the edge of the neck straight, the horns

[1] Ewart, "On the Skulls of Oxen from the Roman Military Station at Newstead, Melrose," *Proc. Zool. Soc.* 1911.

[2] Lydekker subsequently expressed himself as doubtful about the specific validity of this animal. See *The Ox and its Kindred*, London, 1912.

very large and long, near the roots directed outward, and somewhat backward; in the middle they are bent forward, and
towards the points turned a little upward. This colossal species
of Ox, to judge from the skeleton, resembles almost the tame
Ox in form and the proportions of its body; but in its bulk it
is far larger. To judge from the magnitude of its horn-cores it
had much larger horns, even larger than the long-horned breed
of cattle found in the Campania of Rome. According to all
accounts, the colour of this Ox was black; it had white horns,
with long black points; the hide was covered with hair like the
tame Ox, but it was shorter and smooth, with the exception of
the forehead, where it was long and curly[1]."

Fig. 47.   Skull of Urus.   (From Lydekker.)

McKenny Hughes[2] has described the Urus as "a large,
gaunt beast with a long, narrow face."   Fleming[3] mentions a
skull in his possession which was $27\frac{1}{2}$ inches in length and
$11\frac{1}{2}$ inches across the orbits, and Owen[4] refers to a skull which

[1] Nilsson, "On the Extinct and Existing Bovine Animals of Scandinavia,"
Annals and Mag. of Nat. Hist. vol. IV, 2nd series, 1849.

[2] Hughes, "On the most Important Breeds of British Cattle," *Archaeo-
logia*, vol. LV, 1896.

[3] Fleming, *History of British Animals*, London, 1828.

[4] Owen, *British Fossil Mammals*, London, 1846.

was a yard in length.　That the Urus was of great size is proved abundantly; nevertheless Caesar's account of its dimensions must be regarded as much exaggerated.　The following is his description of the animals: "In size they are a trifle smaller than elephants; in kind, colour, and shape they are bulls. Great is their strength and great their speed; nor, having espied them, do they spare either men or beasts.　They are sedulously captured in pits and slain; the young men hardening themselves by such toil and training themselves by this kind of sport; and they who have killed most Uri, proclaimed as such by the horns being exhibited in public, receive great commendation.　But it is not possible to accustom the Uri to men or to tame them, not even though they are caught young.　Their horns differ much in size, shape, and kind from those of our cattle[1]."

Bos frontosus, so named by Nilsson because it had a mesial elevation on the forehead, was in other respects intermediate between *B. longifrons* and *B. primigenius*.　Its remains have been found in Britain.　A local variety of *B. primigenius*, found fossil in Tunis and Algeria, has been called *B. mauritanicus*.　It had a relatively short forehead and somewhat slender limbs. Yet another sub-species of the Urus, called *B. minutus* (since it was a dwarf variety), has been discovered in the superficial deposits of Belgium in association with the Mammoth.

Bos longifrons, which is identified with the Celtic Shorthorn, is the second species of Ox found in Britain, Ireland and various parts of the continent during the Neolithic Age.　According to Hughes it is the only breed which the Romans found indigenous to Britain when they first came there.　It has been found fossil in numerous ancient British and Roman refuse heaps and rubbish pits, as well as in the Swiss Lake Dwellings and in other deposits of similar age.　It was undoubtedly a domestic variety. *Bos brachyceros*, remains of which have been found at Anau in Turkestan, appears to have been the same as *B. longifrons*.　At the present day it is probably represented in its purest form by the Kerry cattle.　*Bos longifrons* has been thus described by Owen: "This small but ancient species or variety of Ox

---

[1] Caesar, *Gallic War*, book vi.

belongs, like our present cattle, to the sub-genus *Bos*, as is shown
by the form of the forehead, and by the origin of the horns from
the extremities of the occipital ridge; but it differs from the
contemporary *Bos primigenius*, not only by its great inferiority
of size, being smaller than the ordinary breeds of domestic
cattle, but also by the horns being proportionately much smaller
and shorter, as well as differently directed, and by the forehead
being less concave. It is, indeed, usually flat; and the frontal
bones extend further beyond the orbits, before they join the
nasal bones, than in *Bos primigenius*. The horn-cores of the
Bos longifrons describe a single short curve outwards and forwards in the plane of the forehead, rarely rising above that
plane, more rarely sinking below it; the cores have a very
rugged exterior, and are usually a little flattened at their upper
part[1]." McKenny Hughes says : "*Bos longifrons* was a very small
animal; probably not larger than a Kerry cow. It was
remarkable for the height of its forehead above its orbits, for
its strongly developed occipital region, and its small horns
curved inward and forward[2]."

It has been shown that *Bos primigenius* probably arose from
Bos planifrons in the East and that *Bos namadicus* was an
Oriental variety of *B. primigenius*. There can be little doubt
that *Bos longifrons* was derived originally from the same source,
and was introduced into Britain by Neolithic farmers and
herdsmen. Duerst[3] has collected evidence of some importance
concerning the history of *Bos longifrons* in Turkestan and Meso-
potamia. He says that in both these localities unfavourable
conditions of life converted "the originally large and stately
Ox" (*Bos macroceros*), which was longhorned and was derived
from the Asiatic Urus (*B. namadicus*), into a stunted, shorthorned form, the *Bos brachyceros* or *longifrons*. The first
remains of *Bos macroceros* at Anau represent oxen which lived
about 8000 B.C. By about 6000 B.C. this animal had become
replaced by *B. brachyceros*. Thus the longhorned variety of

---

[1] Owen, *loc. cit.*
[2] Hughes, *loc. cit.*
[3] Duerst, "Animal Remains from Excavations at Anau," in Pumpelly's
Exploration in Turkestan. Carnegie Institution of Washington, vol. II, 1908.

the early Babylonians gave rise to the small shorthorned breed of Assyrian and later times. Duerst says further that there is reason to believe that the Ox of Anau, which was undergoing this change, eventually reached Central Europe after migrating through Southern Russia and Eastern Europe, while, as already mentioned, it may be supposed to have been transported to Britain by the Neoliths. The longhorned breed, however, survived without much alteration in Babylonia and Egypt, since there is evidence of its existence in these countries about 4000 to 3000 B.C., as well as in India and China, whence it was spread by tribal migrations.

It remains to consider the origin of certain of our modern breeds of cattle and the different views which have been put forward as to their evolution and history. This subject can only be dealt with very briefly, for the evidence which has been presented is chiefly historical rather than geological.

According to some authorities the "wild cattle" now living in Great Britain, that is to say the Chillingham, Cadzow and Chartley breeds, are the descendants of the Urus which was domesticated on the continent in the Neolithic Age and transported to Britain by the English or by the Scandinavian Vikings. This is the view put forward by Boyd Dawkins[1], who points out that there is no evidence of any large domesticated cattle existing in Britain before the arrival of the English. On the other hand it has been suggested that the Urus may have survived in Scotland in sufficient numbers to have given rise to the Atholl and other Scottish "wild" cattle, but there is no evidence in favour of such a view. For although the Urus was widely distributed in Great Britain in the Neolithic period it would appear probably to have become extinct (at least in England) by the time of the Roman invasion.

Such considerations have led McKenny Hughes to believe that *Bos primigenius* has had no share whatever in forming either the Chillingham or any other British breeds of cattle. This geologist supposes the Chillingham breed to be descended from cattle imported by the Romans, pointing out that there

---

[1] Dawkins, "Chartley White Cattle," *Trans. North Staffordshire Field Club*, vol. xxxiii, 1899.

is a close resemblance between the Chillinghams as they exist to-day and certain modern Italian cattle. He derives further evidence from contemporary drawings of Roman and Ancient Egyptian cattle. Against this view, which has been adopted by James Wilson[1], it has been pointed out that there is no evidence that the Romans imported cattle into Britain, that it is unlikely that they would have done so, having regard to the fact that domestic cattle already existed in Britain, and in view of the difficulties of transporting animals in sufficient numbers to have given rise to English breeds.

Hughes has supposed further that the imported Roman breed blended with a certain admixture of Celtic Shorthorn (already present in the country) to give rise to the Highland and Welsh cattle. The mediaeval Shorthorn is regarded as a reversion towards the native variety after the withdrawal of the Roman legionaries when cattle breeding was no longer carried on by the same careful selection as formerly. Lastly, the longhorned breeds Hughes shows probably to be descended from Holstein cattle imported in the Middle Ages[2].

Cossar Ewart[3] has shown that in the Roman camp at New-stead the remains of five fairly distinct types of oxen occur in addition to animals that were probably crossbred. These five types are as follows: (1) The Celtic Shorthorn. (2) A longhorned type allied to the Urus. Such animals on Hughes's theory were presumably imported by the Romans. (3) Oxen with deep notches below the horn-cores and with the hind view of the skull generally bearing a close resemblance to *Bos acutifrons* of the Siwalik Pliocene. It is possible that this type may have been of the nature of a reversion, brought to the surface by cross-breeding, since there are some reasons for supposing that *B. acutifrons* was ancestral to *B. primigenius* or at least to *B. namadicus*. (4) Oxen with a convex forehead and horns curving backwards and downwards, and with some affinities to *B. namadicus*. These also may in certain characters have been reversionary, since there is no evidence that *Bos namadicus*

[1] Wilson, *The Evolution of British Cattle*, London, 1909.
[2] Hughes, *loc. cit.*
[3] Ewart, *loc. cit.*

extended westwards. (5) Hornless cattle, some of which had
a mesial prominence and therefore belonged to the *frontosus*
type of Nilsson.

The question as to the origin of hornless cattle is one to
which no answer can be given. Ewart has pointed out that
such cattle existed in Egypt under the Fourth Dynasty, and
that according to Tacitus a hornless breed lived in Germany
in the first century. It has already been mentioned that
species of *Leptobos* living in Pliocene times were almost certainly
hornless in the female. It may be that a reduction in the horns
took place gradually, or it may have occurred suddenly through
a "mutation." The disappearance of horns may have taken
place again and again in the developmental progress of the
different breeds, and so may be regarded (in certain cases at
least) as of the nature of a reversion to a previously existing
type. These are problems concerning which there is little
evidence forthcoming towards a solution. There are, however,
records of horned cattle producing hornless "sports." James
C. Lyell[1] has put forward the theory that the Polled Angus
Cattle were brought from Norway, and James Wilson[2] has
contended that all the hornless breeds of Britain have had a
Scandinavian origin.

It will be seen from the foregoing account that whereas the
origin of the various types of domestic cattle is still a subject
of much controversy, the available evidence points to the conclusion that they were all derived in the first instance from
some species identical with or closely similar to the wild Urus
or *B. primigenius*, which was probably first domesticated somewhere in Asia. The humped cattle of India (*Bos indicus*)
however apparently had a different origin, being descended from
an animal resembling the Javan Banting (*Bos sondaicus*),
while the Gayal or domesticated Ox of certain Indo-Malay and
Indo-Chinese races is not markedly different from the wild
Gaur (*B. gaurus*) from which it is probably derived[3]. The
history of these species and of the domesticated Buffalo lies
outside the scope of this volume.

[1] Lyell, *The Polled Cattle of Angus*, 1882.   [2] Wilson, *loc. cit.*
[3] Lydekker, *loc. cit.*

THE SHEEP

The tendency of opinion among zoologists in recent years has been towards assuming that the Domesticated Sheep, like the Horse, has had a multiple origin, and that of the four main types of wild sheep now in existence, namely the Mouflon, the Urial, the Argali, and the Bighorn, two if not three have been concerned in the ancestry of the domesticated breeds. Moreover, there are strong reasons for supposing that, as in the case of the Ox, the process of domestication first took place in Asia, the domesticated varieties being introduced by the ancestors of the Swiss Lake Dwellers into Europe, where they have since undergone great variation by cross-breeding.

The origin of the genus *Ovis* in geological times is a problem that is still more obscure, and fossil remains of sheep-like animals have done little to elucidate it. The Pliocene *Criotherium* from the Isle of Samos, an animal with strange twisted horns, had certain affinities to the Sheep, but it is not supposed to have been in the direct line of descent. A similar statement may be made about the extinct antelopes of the genus *Oioceros* which have been found fossil in the Lower Pliocene of Greece, and had spiral horns twisted after the manner of sheep.

Coming to more recent times a portion of the skull of a sheep has been found in the Cromer Forest Bed. In this animal the curvature of the horns was like that of the Mouflon, but certain other characters, and more particularly the size, are suggestive of an animal allied to the Urial. The Forest Bed Sheep has been called *O. savini*. It lived in Britain contemporaneously with the Elephant and the Rhinoceros, and then died out.

The Argali type was represented in Pleistocene times by *O. antiqua* (in the south of France), *O. argaloides* (in Moravia) and *O. ammon fossilis* (in Transbaikalia).

Sheep belonging to the Bighorn and Arui types have also been found fossil in Pleistocene times.

Of the types now in existence the Bighorn of America has never been domesticated, and the Arui of the same continent is an aberrant species which has sometimes been referred to a separate genus.

It remains therefore to consider the Mouflon, the Urial, and the Argali types, all of which have been supposed to have had a share in giving rise to one or other of our domesticated varieties.

The Mouflon at present inhabits certain parts of western Asia and various islands in the Mediterranean. One species (*O. musimon*) inhabits Corsica and Sardinia. Another closely

Fig. 48.   Skull of Sardinian Mouflon Ram with normal horns.   (From Ewart.)

allied but smaller species (*O. orientalis*) lives in Cyprus, and there are several related species or varieties in south-west Asia. In general appearance the Mouflon is more antelope-like than any domestic sheep, and the resemblance, which is due largely to its light, neat build, is increased further by its coat of short hair. The tail is short and the head comparatively long with a flat forehead, and in the male large spirally coiled horns with their tips, in normal individuals, directed forwards. The limbs are long and slender and terminate in sharp-edged hoofs.

The Urial (*O. vignei*) is found in Turkestan and Afghanistan and the neighbouring countries as far east as the Punjab and Tibet, where it lives at an elevation of 14,000 feet. The skeleton is similar to that of the Mouflon, but the large and deep pit on the front of the orbit, where the face gland is situated, is characteristic of the Urial.   The horns form a very close spiral. In the female there are small upright goat-like horns.

The Argali (*O. ammon*) is the largest of all the wild sheep, being sometimes as high as 13 hands at the withers.   It is found mainly on the mountain ranges around the Gobi deserts, and the largest variety inhabits the region of the Great Altai Mountains.   The horns are frequently extraordinarily massive and spirally coiled as in the Mouflon.   In typical specimens the horns of the old rams form more than one complete circle, their tips reaching considerably beyond the lateral margins, while in front they are rounded and almost touch the sides of the face.   They may reach a length of 62 inches and a girth of

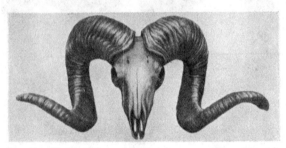

Fig. 49.   Skull and horns of Altai Ammon.   (From Ewart.)

20 inches.   In some kinds of Ammon the horns are "nipped in" like those of the Scottish Blackfaced Sheep, while the horns of other varieties are like those of Merino rams.

It is now generally admitted that both the Mouflon and the Urial have taken part in forming some of the breeds of domestic sheep.   The half wild sheep of Soay show a resemblance to both these types, sometimes to one type much more than the other.   The Shetland variety of peat sheep and the sheep whose remains were found in the Roman camp at Newstead have been shown by Ewart[1] to be of the Urial type.   Moreover the Turbary sheep (*O. aries palustris*) of the oldest Swiss Lake Dwellings were essentially Urials like the sheep domesticated

[1] Ewart, "Domestic Sheep and their Wild Ancestors," *Trans. Highland and Agric. Soc.* 1913 and 1914.

by the Anauli in Turkestan, whence they were carried
westwards. Elwes[1] believes that the diminutive sheep still
existing in the Orkneys are probably derived from the Turbary
Sheep. According to Lydekker[2] there is nothing in the physical
characteristics "to preclude all the British long-tailed sheep
being descended from the Mouflon, all the horned breeds having
horns of the general type of those of the Mouflon," but he
admits the probability of a considerable admixture of Urial
blood in the British domesticated breeds.

Fig. 50. Skull and horn-cores of Bronze Age Sheep, according to Ewart by
   origin partly Ammon. (From a photograph by Prof. Ewart.)

Ewart is of opinion that in addition to the Mouflon and the
Urial, the Argali sheep have had a share in forming certain
breeds, and in support of this view he cites the work of Douglas
Carruthers[3], who has shown that the tribes of Central Asia,
both in ancient and modern times, have constantly crossed

[1] Elwes, *Scottish Naturalist*, 1912.
[2] Lydekker, *The Sheep and its Cousins*. London, 1912.
[3] Carruthers, *Unknown Mongolia*, London, 1913.

their domesticated ewes with wild rams of the Argali or Ammon type. Ewart adduces further evidence based on the examination of remains from Pleistocene deposits, on a comparison between the skeletons of wild species and primitive and modern breeds, and on cross-breeding experiments, in support of the view that the Scottish Blackfaced Sheep, the Merino, and possibly other British breeds, are in part descended from the Argali[1].

After the beginning of the Ice Age, when *O. savini* became extinct, there were no sheep in Britain. They reappeared with the coming of the Neoliths, who, as we have already seen, brought with them their domestic animals, and among these

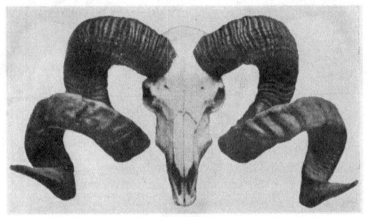

Fig. 51.   Skull of Dorset Ram showing horns of Ammon type.   (From Ewart.)

were sheep of the Mouflon and Urial types. Moreover, Ewart has shown that some of their sheep had horns like those of a small Ammon, and he points out further that the remains of sheep found in the alluvium of the Thames Valley show a marked resemblance to the Argali type.

THE PIG

The family to which the Pig belongs is the least altered or most primitive of the hoofed animals. They differ from horses, sheep and oxen in having the ulna of the forelimb and the

[1] Ewart, *loc. cit.*

fibula of the hindlimb as separate bones, distinct from the radius
and the tibia. The bones of the wrist and ankle (carpals and
tarsals) are not fused together. The molar teeth are bunodont,
that is to say their crowns bear simple or roundish bosses or
tubercles, and not crescent shaped structures as in Ruminants,
or complicated irregular ridges as in horses. The canine teeth
are well developed and the complete mammalian dentition of
forty-four teeth is generally present.

We find the above-mentioned characters in *Homacodon* of
the Middle Eocene of Wyoming, U.S.A. This animal was no
larger than a rabbit and had five digits, but it was essentially a
primitive pig. *Cheropotamus*, of the Upper Eocene of the Isle
of Wight and France, was larger but not very dissimilar. The
Miocene *Hyotherium* which ranged over Europe and Asia was
as large as the Wild Boar. *Listriodon*, which had a similar wide
range, was remarkable in having the cusps of the molar teeth
fused so as to form complete transverse ridges like those of the
Tapir. *Elotherium*, which has been found in Upper Eocene or
Miocene strata both in Europe and North America, was a more
specialized form. It had two toes only and the limbs were long
and slender. The brain was remarkably small. *Hippohyus*
was a Lower Pliocene pig found fossil in the Siwaliks. It was
very like *Sus* but had the cusps of the molars modified in
irregularly arranged laminae.

Coming to the genus *Sus*, to which the Domestic Pig belongs,
we find it first in the Middle Miocene of France and Italy. This
species, which is called *S. choeroides*, had simple molars like
those of the existing *S. andamanensis*. *Sus erymanthius*, or
the Erymanthian Boar, has been found in the Red Crag of
Suffolk and in various continental countries. The still larger
S. titan and *S. giganteus* occur fossil in the Lower Pliocene of
the Siwalik Hills[1]. There is no evidence, however, that any
of these or of the other fossil species of *Sus* which have been
described took part in the formation of *Sus scrofa*, which
includes the Wild Boar of Europe, and all the various breeds of
domesticated pigs, excepting certain Oriental varieties which
may have had a different origin.

[1] See Smith Woodward, *Vertebrate Palaeontology*, Cambridge, 1898.

Sus scrofa itself occurs fossil in the Upper Pliocene of Europe. In England it has been found in the Cromer Forest Bed, and it survived until the seventeenth century, when it was still abundant in Ireland. Its remains have been found also in numerous Pleistocene deposits, such as those of the Cambridge-shire fens.

Rutimeyer and Nathusius and some other naturalists have supposed that the various breeds of domestic pigs have been descended from two or more wild species, one of which was supposed to be the same as *Sus scrofa* while another was an Oriental species or variety called by Pallas *S. indica*. More recently a considerable number of other kinds of Oriental pigs have been described (e.g. *Sus papuensis, S. timorensis, S. anda-manensis, S. leucomystax*, etc.), but it is doubtful how far these are really distinct. Moreover it has been suggested that certain supposed species of pigs really represent animals which have become feral, and there can be no doubt that the wild or semi-wild pigs of certain islands in the New World belong to this category.

It is probably safe to assume that all the British breeds of the Domestic Pig are the descendants of *Sus scrofa*, which, as we have just seen, lived in Britain continuously from Pliocene until quite recent times. On the other hand it is believed that the Neoliths brought domestic animals with them from the East, and among these were the Turbary Sheep, the *longifrons* variety of the Ox, as well as the Dog and the Pig. It is possible therefore that the last named animal may have been in part descended from some wild species other than *Sus scrofa*; and further that the Domesticated Pig of the Neoliths may have been the part ancestor of some of our modern breeds. Such suggestions are however extremely speculative.

At the present day the domesticated breeds of China and Siam have broader and stouter heads than one would be inclined to expect in descendants of *Sus scrofa*. It is very probable that these varieties are descended from Oriental species other than the Wild Boar of Europe.

However this may be, there are some breeds, such as the old Irish Greyhound Pigs, which are almost certainly the pure descendants of *S. scrofa*, since they resemble it in the length of their legs, the development of their canine teeth into tusks, and the comparative thickness of their hair. Concerning the origin of certain aberrant varieties, such as the solid-hoofed pigs which were known to Aristotle and exist at the present time in America, it is impossible even to guess.

THE DOG

The family *Canidae*, to which the Dog, the Wolf and the Fox belong, first appears in the Upper Eocene of Europe, where it is represented by the genus *Cynodictis*. This animal, which is probably the ancestor of *Canis*, and possibly also of the weasel family, was very like the Dog in its skeletal characters, one of the most noticeable differences being the expanded end of the humerus. It had five well-developed toes on each foot.

Canine animals were common in Miocene times in both Europe and North America, but the only complete skeletons discovered are those of the so-called "Fossil Fox," *Galecynus oeningensis*, from the Upper Miocene of Baden, and an allied species, *G. geismarianus*, from strata of the same age in Oregon, U.S.A. In the European skeleton the first digit of the forelimb is larger than in the Dog, and in the American skeleton the humerus resembles *Cynodictis*, to which *Galecynus* was obviously allied[1].

The *Canidae* appear to have reached both India and South America in Pliocene times. Subsequently they arrived in Australia, whither they were not improbably carried by Man.

The genus *Canis* first appears in the Cromer Forest Bed, where it is represented by remains of the Wolf and the Fox.

The Domestic Dog (*Canis familiaris*) was probably the first animal tamed by Man, but nothing is definitely known about its origin. Its remains have been found in the Danish kitchen-middens, and in the Lake Dwellings of Switzerland, and various deposits of contemporaneous age. As already mentioned, there is evidence that the Neoliths in travelling westwards brought with them a Domesticated Dog from the East. The

[1] See Smith Woodward, *loc. cit.*

animal was on an average about the size of a Sheep Dog.   In
the Bronze Age a larger Dog existed and in the Iron Age there
was a still larger variety.   The Jackal (of which there are several
species), the Indian Wolf (*C. pallipes*), and the Bunasu (*C. pri-
maevus*) have all been suggested as ancestors of the Dog.   It is
possible that there were really several ancestors which are
represented in various proportions in the two hundred or more
varieties which exist at the present day.   In the absence of
more evidence a further discussion of the problem as to the
origin of the Domesticated Dog would be unprofitable.   It may
be pointed out, however, that physiologically there is no
specific distinction to be drawn between the Dog and the Wolf
and certain other wild species of *Canis*, since these animals will
generally breed together in confinement and produce offspring
which are fertile.

INDEX

Printed in the United States
By Bookmasters